LOCOMOTIVE AP

At the North British Locomotive Co.

by

Nigel S. C. Macmillan, M.Sc., C.Eng., M.I.Mech.E.

Plateway Press, P.O. Box 973, Brighton, BN2 2TG

1 871980 10 0

By the same Author

The Campbeltown and Machrihanish Light Railway

N.S.C.M. 15·3·79

© Nigel S. C. Macmillan/Plateway Press 1992

Printed in Great Britain by Wayzgoose plc, East Road, Sleaford, Lincs.

Front cover illustration: East African Railways and Harbours Class 29 2-8-2 on the crane in the North British Loco. Co.'s erecting shop. In the background the valve squad work on an LNER B1 4-6-0. The Author was beside the camera when the picture was taken. (Mitchell Library)

Back cover illustration: NBL advertisement, 1952 (Railway Gazette)

Frontispiece: A spanking new British Railways B1 4-6-0 raises steam outside the NBL paint shop prior to delivery. (Author)

Front cover artwork by Richard Horne; book design by Keith Taylorson.
Line drawings, scale diagrams and maps are by the Author, as are all photographs not credited individually.

CONTENTS

FOREWORD

The author of this book can justly claim to be a born engineer, a fact which will become apparent as soon as you start to read it. His whole life is centred round his railway interests and engineering in general, and is one of the dying breed of engineers who were brought up in the tradition established by such engineering giants as Brunel and Stephenson.

With no precedents to guide them they had perforce to be all-rounders; and the tradition they established flourished until the end of the steam age, when a generation of specialists succeeded them.

Nigel Macmillan was fortunate in that his engineering training was acquired at a period when the old traditions, although soon to disappear, were still alive; and he was able to benefit thereby. Thus the period described by him as regards locomotive engineering, although in the 20th century, really represents the final fling of 19th century technology. This book, therefore, is an account, by an eye-witness, of what went on behind the walls of the engineering shops of Glasgow and which resulted in the creation of the splendidly built and painted locomotives, the few preserved survivors of which arouse the utmost enthusiasm whenever they appear in public.

I first met Nigel Macmillan many years ago through a common interest in the Campbeltown & Machrihanish Light Railway, Scotland's only public narrow-gauge line. He could claim to be an expert on this little line, situated far away from the main centres of population in the remote Kintyre peninsula in the West of Scotland. Through his help I was enabled to model it complete.

All engineers have to be trained, so Nigel became a railway apprentice and this book is, therefore, an account of those early days. It contains a great deal of vividly expressed information (now of great historical value) about railway locomotive construction in the late 1940s and early 1950s as well as a good number of anecdotes about the men, including himself, engaged on this work, and some personal anecdotes, amusing and, in some cases, hair-raising; especially the story of how he was, on one occasion, chased by a rake of runaway wagons.

Towards the end of this book you will learn the reasons which impelled him to abandon this career. From being the master of a dead technology, he acquired fresh proficiency in machine tools, coal mining, and offshore oil; but his memories of his early days are sharp and vivid and he has done all railway historians a great service by committing those memories to paper.

D. A. BOREHAM
Vice President
The Model Railway Club

NORTH BRITISH LOCOMOTIVE
COMPANY Lᵀᴰ. GLASGOW
N° 28000

PART I

Chapter 1

GIFFNOCK QUARRIES

Two things stand out about my life – firstly I started my apprenticeship in locomotive engineering when I was 6 years old; secondly I served the formal part of my 'time' in the middle of the nineteenth century. These apparently absurd comments will become more reasonable as the story unfolds.

One of the highest amenities of the house in which I was raised, was the Glasgow to East Kilbride railway line on a high embankment at the foot of the garden. The line climbs steadily at 1 in 70 from Glasgow. Trains from the city had a stiff start from Giffnock station about half a mile away and all but the best engines laboured up the line behind the house; but trains to town rattled down at a fine old rate. Passenger trains were drawn by ex-Caledonian Railway 4-4-0's, L.M.S. class 2P 4-4-0's, C.R. 0-4-4 tanks and the huge C.R. 4-6-2 tank engines generally known as Wemyss Bay Tanks. These 'big pugs', as the drivers called them made a great deal of noise as they emerged from a cutting on to the long embankment behind the house. As the exhaust beat echoed back and forth, it fitted in exactly with the middle bars of Tchaikovski's 'Dance of the Mirlitons' well known in our house. I never hear the piece now without being immediately transported back and seeing one of these big engines thrashing up the hill. By that time they had been largely displaced from the coast lines by Fowler and later Stanier 2-6-4 tanks, some of which appeared occasionally on our line. Freight was almost entirely in the hands of Caley 0-6-0 tender engines.

A few hundred yards up the line from the house, the East Kilbride line crossed the Caley's Lanarkshire and Ayrshire line which carried a passenger service to Uplawmoor and was worked by a similar collection of engines. This line in those days still carried a few boat trains to Ardrossan which had more exotic engines, like 4-4-0 compounds on them.

One working which didn't make much sense was a little tank engine, later to be known as a Caley 3F 0-6-0 T, which would wander up the line to Clarkston, the next station, reverse over the crossover there and trundle back down again. Once it appeared piloting a failed Wemyss Bay Tank and managed to lift the big engine and its five coaches up the hill with a struggle. Later knowledge brought the realisation that the train engine would be needed to work the brakes as few 3F's had vacuum fittings.

Primary school was twenty minutes' walk from home but adjacent to Giffnock station. The longer way home was over the station footbridge and it was found that the 3F could often be seen shunting in the yard there. On one occasion, after a pleasant half hour had been spent watching the engine playing with its trucks, it came to rest level with me with only a rickety fence between it and the footpath. Even a small locomotive is a remarkably big thing when viewed from track level, even more so to a small boy. "Want to come up, son?" the driver shouted down and I was through the fence like a shot lest he changed his mind, oblivious of potential dirt on school clothes. First impressions of a steam locomotive cab are steel all round, heat, small coal getting trodden into the wooden floor. The boiler backplate with its gauges and levers had a magic look and the whole thing was dominated by the open firebox door while the fireman made up the fire.

Without saying any more, the driver reached for the whistle and opened the regulator. A whole Hell of noise broke loose as the engine lost her footing and slipped, gripped and slipped again pushing her wagons in front of her.

The next move was to propel her train of wagons loaded with ash into what had always been taken to be a siding and now turned out to be a long rickety track leaving the station far behind. This was my first footplate run and started my training. It was a surprise to a small boy to find that a locomotive's 'puff' sounded so different from the inside, that the entire engine jarred and rattled with every beat and the fire flared and tore at the firehole in unison as the fireman put coal in.

This was the Giffnock Quarry branch, owned by William Beardmore & Co. Ltd., and used at this period,

1. Ex-Caledonian Railway 0-6-0T shunts wagons on Beardmores' coup line at Giffnock Quarries in June 1959.

2. The result of the runaway described in the text. Standard mineral wagons in Giffnock Quarry, August 1960.

to dispose of ash from their Parkhead Forge. Half of Glasgow had been built with sandstone from these quarries. Now they were more or less exhausted, urban development had spread all round, and houses were now built of brick and rough-cast. The 3F brought a daily train of about 20 mineral wagons of ash and deposited them on the edge of whichever quarry suited the bill and took the previous days' empties away. As the branch trailed on to the main line through the goods yard above the two station crossovers, it was easier for the engine to pop up to Clarkston and back to round its train, hence the light engine working I had observed. There was an ancient water tower on the branch line but this was often dry and sometimes the engine had to go further up the line to Busby for water. Beardmore's had been tipping ash into the quarries since 1913 and kept an engine of their own there around 1918. This was an Avonside 0-4-0ST (1779 of 1917) and it is possible that another Parkhead Forge engine worked there about the same time.

In these days, there was still a trace of the line into the quarries, the points were still in situ but the rest lifted, leaving the inevitable sleeper marks. This went under a minor road by a stone bridge; the parapets still show near where Beardmore's line crossed the same road by a gated level crossing. Immediately after this crossing was a spring catch point with a removable lever followed by a loop, one leg of which seemed to be full of wagons. The line then climbed up to Braidbar farm where there was a reversal and the wagons were run down the other side of one quarry and left to be emptied. This was achieved by men with shovels, and there was always a parapet of heaps of bricks and ash until the rails were moved forward each weekend. The quarries were deep and filled with water and very dangerous. Going anywhere near them was absolutely fobidden and once, when about 6 years old, I fell in and had great difficulty explaining how I got soaked to the skin (I was wearing a kilt and this could not be dried in a hurry). The quarries were also extensive. Twenty years later my wife and I carried an 8-foot pram dinghy to them and spent the day exploring, using an outboard motor.

During the war, the ash in Braidbar quarry started to burn. It burnt for years, so fiercely at times that the track had to be lifted as all the sleepers charred away and tipping activity concentrated on the New Braidbar quarry on the other side of the hill. Water from this latter was pumped through a pipe laid alongside the railway and channelled through trenches dug all over the surface. Trees that had seeded on older tippings gradually went brown and died. It must have been ten years before the fire was finally extinguished and its acrid smoke, once smelt never forgotten, no longer eddied over the neighbourhood.

Whether the ash itself was supporting combustion or old coal workings from the nearby Giffnock Colliery (closed 1927) was never cleared up. But underground fire broke out at the site of the colliery just as it was being conquered at the quarries and burned there for some years more.

For some reason, throughout the history of the line the temporary tip line always managed to slope steeply down to the dead end. In Scotland, this type of tip is always referred to as a 'coup' (pronounced as in 'cow'). Sometimes more than the contents of the wagons were 'couped', the whole rake would get away from the engine and most of the wagons would go over the end. I first encountered this first hand one Saturday. There were then stables on adjacent ground to the quarry and the line had advanced to just such a degree that the stops, a loose pile of sleepers, were in line with the buildings. My parents had business at the stables and this day I was with them. I could watch the 3F shunting officially.

Leaning on my bicycle beside the stable, I watched the wagons coming towards me and was surprised when the men involved shouted to me to get away. Now, I had every right to be there. Wasn't I with my parents? So I stood my ground and waited to see what happened when the wagons hit the sleepers. They did not even hesitate. When I saw the pile of sleepers topple and the leading wagon vault over, I realised things were no longer under control and I dropped my bike and ran. I never ran so hard in my life. Behind me there was the most horrendous of noises and the train ran into the stables, a large wooden building. The first wagon entered the building through the wall, tipped up on end and came through the roof and the rest of the wagons piled into it. I ran until my way was blocked by the fence round the exercise yard. I did consider jumping it and probably would have done just that but the noise had stopped. There was a huge pall of dust and people appeared from all directions. By the time I got to the crowd surrounding the wreckage, I was on the outside and two men were frantically trying to pull the mangled remains of my bicycle from under a wagon. They were also looking for the mangled remains of *me* until someone spotted me. Inside the stable, the corner stall had been empty but was now full of smashed timber and broken wagonry. The owner of the place had been grooming a horse in the next stall and was able to hold its head through the commotion. He was almost stone deaf and this probably saved him. They had to take the remaining stall partition down to free him and the horse. Miraculously, no one suffered more than shock and the building was eventually repaired. However, a large sliding door at the other end never ran on its bottom rails again.

This was the most spectacular runaway, but they continued right through the life of the line and the end of the tips always had an assortment of W-irons, solebars, etc., as mute testimony. The last smash-up took place in 1959 when some boys released the brakes on a Sunday and let a rake go. I lived nearby at this time and heard the remaining brakes squealing as the wagons moved. I just got to the door in time to see the lot going over the end for the second time in my life, though this time from 400 yards away. By now the wagons were MOT 16 ton steel ones and they made a noise like a thousand dustbins falling.

Shortly after this the operation stopped but an enormous dragline excavator was called in and it spent the next few years digging over the ash extracting scrap metal from it; and a not inconsiderable amount was bits of wagons.

The engine on the spoil train was almost invariably a C.R. 3F 0-6-0T. Only once did I see Midland style 0-6-0T. The driver called it an Irish engine. Goodness knows why. Maybe he saw one on the N.C.C. in Belfast. And another time, much later on, a B.R. standard class 5 turned up light with jacks all over its front platform to rescue the tip engine which was trapped behind a derailed brake van. And of course the 3F eventually gave way to an English Electric side rod diesel 0-6-0, but that was near the end.

3.   Successor to the CR 0-6-0T, an English Electric 0-6-0 diesel gingerly drops wagons down the line towards the stops on the Giffnock quarry line. 11 June 1961.

Chapter 2

CLARKSTON CONNECTIONS

Between Giffnock and Clarkston the East Kilbride branch crosses over the Neilston electric line by an overbridge. The upper line was first on the scene, being built from Busby Junction to Busby in 1866 and later extended through to the village of East Kilbride and thence on to join the Strathaven (pronounced 'Straven') line at Hunthill junction in High Blantyre. Initially it was single track from Giffnock onwards, but in 1881 it was doubled as far as Busby but to this day it remains single thereafter.

During the nineteenth century, the Glasgow and South Western Railway had a monopoly of traffic, mainly coal, to all the Ayrshire ports. To compete with the G&SW, the Caledonian built its own line to Ardrossan from its Beith branch off the Glasgow Barrhead and Kilmarnock joint line. This new line was called the Lanarkshire & Ayrshire, and was nominally a separate company. It was then extended back towards Glasgow in stages and where it burrowed under the East Kilbride branch two curves were laid in to allow traffic access from one line to the other. As both lines were climbing steeply, the curve between Clarkston and the site of the present Williamwood station was fairly level albeit on a high embankment and requiring a bridge over a main road. The other curve had to climb away from the already steep grade of the L&A and took well over a mile to gain sufficient height to join the other line. The first curve had a double track laid on it and some coal trains were run over between 1903 and 1904 when it was officially closed. Later Williamwood Halt was built on the site of the Clarkston West Junction. For some reason it was found more convenient to route traffic through connections nearer the city.

The fate of the other curve as a through line is even more nebulous although the track at least remained in situ until 1962. The Board of Trade apparently did not take to the idea of two facing junctions on a 1 in 70 falling gradient and refused permission to use the Clarkston/Muirend link. It is not clear if the rails were actually laid in but the writer inclines to think not. The two tracks ended, not in conventional buffer stops, but in two 2ft 0in square beams bolted to the rails, and sleepers and chairs without rails went almost to the other line; as if the construction gang had just gone home all those years ago.

The tracks were soon lifted on the west curve but the east connection was used for the next sixty years as stock sidings. Each of the two tracks could accommodate six 10-coach trains and when travelling into Glasgow from Williamwood, these carriages were a familiar sight, appearing first far over a patch of waste ground, and then alongside high up above the train. As Muirend drew near, they gradually got lower and lower until the last were on the same level. These sets tended to be older corridor stock which made them particularly interesting. Even green Southern and brown & cream Great Western stock managed to find its way there.

During my early childhood, I never saw anything move on the sidings; they must have gone by night mostly. But one day early in the war, I heard from the garden an engine slipping and spluttering in the direction of the sidings. On closer investigation (one had to trespass, there was no other way there) it turned out to be an LMS type Stanier 2-8-0 propelling a set of 10 corridor coaches up to the end of the sidings. It was an odd engine. To start with it was painted light brown; it had a Westinghouse pump on the smokebox and it had W↑D on the tender. When it finally made it to the top end of the line, four cast-iron wedges were hauled out of the brake van and placed over the rails on the downhill side of the end coach bogie's wheels. The 2-8-0 then pulled the train up on top of these wedges before uncoupling and ambling light back to Muirend. I went with it – my first footplate ride in a big engine. At Muirend I had to get off before the main line was joined lest the station master should see a schoolboy on the footplate.

From then on, all that summer, and indeed for many years, every time carriages came up the sidings I was there. The engines were usually LMS 2-8-0's or WD Austerity 2-8-0's with the occasional Caley 60 class 4-6-0. After that first one, I never saw another WD Stanier. There was even a Compound 4-4-0 one day and it took hours to propel its set to the end and gave up with about four coach lengths to go. That was all right until the sixth set sent up on the same line wouldn't fit and there was no way its engine was going to propel 60 coaches up *that* gradient!

4.   An ex CR Jumbo 0-6-0 is wreathed in steam from a leaking injector overflow as it waits to collect stored wagons. The train crew have abandoned the engine to close the side doors on some 60 wagons. January 1969.

5.   The bridge carrying the East Kilbride line over the Neilston line being raised to accommodate electrification of the latter.

At the Muirend end of the sidings, the two lines converged to single before joining the main line and access to the other road was by crossover. This is a fairly modern idea instead of 'double junctions' but this was the only place I had seen this form at that time. Trains coming off the sidings were held by two drop front ground signals side by side followed by *two* catch points controlled from the signal box. Going into the sidings, there was a full arm bracket on the starting signal, controlling the movement as there was no advanced starter. In fact, engines rounding their trains proceeded past the starter before coming back through the crossover. The bracket was unusual in having two lower quadrant arms of different types, one typically Caledonian and the other North British. Both these arms are now in the author's possession.

Then the track on the sidings was interesting. On the two adjacent main lines, standard LMS 3-screw chairs were used although on the East Kilbride branch these only had two screws inserted except at rail joints. On the carriage sidings, old CR 4-spike chairs were used, and the rails were joined by Spooner patent fishplates that looked for all the world like giant versions of those used by Bassett-Lowke on their model track. These fishplates were shaped to embrace the lower web of the rail. (This method was employed much later by British Railways in a slightly modified form.) In Muirend station yard, an older form of spike chair was used and the matching check-rail chairs at points carried the letters 'L&A'. The L&A stopped officially at Clarkston West Junction where it joined the Caledonian end on, which makes this rather a puzzle.

The storing of carriages was carried on until the mid '50's when vandalism made it no longer attractive. For some time, cripple mineral wagons were kept there. After a while all the side doors were dropped by the local yobos, and, between the two lines of rails, the doors lay on adjacent wagons. When the time came to remove these wagons, the unfortunate train crew had to close all the wagon drop doors on their own train plus those on one side of the one on the next line. One day they were walking back to the engine between two lines when they came round the bend and found the Apaches working towards them dropping the doors again! Suffice to say, good old-fashioned justice was done.

The last use made of the sidings was to store scores of brand new 20T hopper wagons which had been ordered prematurely. In 1962 the tracks were lifted back towards Muirend leaving two spurs for one football special each. This only lasted a short while before they too were removed along with all the local goods yards under the Beeching rationalising of freight operation.

Haunting Muirend station as I did, always in the evening after the station master had gone off duty, it was natural that I should get to know the station staff and before long I was invited into the signal box. There I

6.   A W.D. 2-8-0 on the connecting line between Muirend and Clarkston, April 1959.

11

was initiated into the peculiar rites of the block system and expressions like "three pause one" (bell description for a local passenger train) and "train entering section" became an unerasable part of my vocabulary. This served me well as, later on, I was to know the signalman in Pollokshaws South and Kennishead signal boxes and to assist in the working of these boxes. It was at Kennishead that I made the terrible faux pas of letting a heavy southbound mineral go through ahead of the London sleeper. I had not realised that Nitshill box was switched out in the evening and the next place he could take refuge was Neilston Low – some miles away and half-way up Neilston bank. The sleeper was offered and accepted and we still hadn't got out of section for the frieght. He came up slowly to our outer home signal pulled by a 5×P 4-6-0 with a 2P 4-4-0 as pilot and we cleared the signal just as he got there. The same thing happened at the inner home and just as he approached the starter and came to a dead stand we got "line clear" and pulled off the board. Starting a heavy sleeper on that grade was no joke – I've never seen an angrier train crew.

Then there was the Carlisle local – all stations to Glasgow. This was usually five coaches and a Compound 4-4-0. The guard always made a great thing of waving off the train, carelessly chatting to the porter, and swinging into his passing van in a nonchalant manner. Highly professional and impressive – until the evening there were only four coaches. He missed it completely. We had to have his train held at Pollokshaws and have the shunting engine there come up 'wrong line' to rescue him! Poor man, forever after his gas was turned down to a peep.

7.   Nearing the end of an era, a BR 2-6-0 waits with a football special at Muirend.

8.   One of many interesting coaches stored over the years on the Muirend/Clarkston link, an ex CR Grampion corridor 12 wheeler. The coach is in LMS livery with the class 3's painted out and SC and M added in white to the LMS number. The coach to the left is a G&SWR corridor brake showing the distinctive large van windows and vertical white handbrake. 23 March 1952.

12

Chapter 3

DUNURE & MAIDENS LIGHT RAILWAY

Towards the end of the Hitler war, I spent some happy summers in the Ayrshire village of Dunure. Lodging was with a Mrs Simpson one of whose sons, Bill, later became famous on television as Dr Findlay, and whose house was not in Dunure village but near the erstwhile G&SW railway station. Although the railway passed close to the village, the station was, as was customary, about a mile away.

On the Monday morning of the first day, as breakfast was finishing, the unmistakable sound of steam safety valves lifting was heard and I was out of the house and up the road in a flash. It turned out to be an LMS class 2P 4-4-0 on a goods train standing at the (closed) station. Dunure station was in a deep cutting and had been, like most on the line, an island platform from which one line had now been lifted. In the goods yard, on level ground above the station, there was a small office from which all business and traffic control was conducted. As this was a telephone post the train had to stop in the cutting and wait until the one and only official came down and gave it permission to go on. While this procedure was taking place, I burst on the scene, slowing to a nonchalant and discreet stroll around the engine but eventually manoeuvring such that I was asked on to the footplate. With the right away came "Just stay on, son, and I'll drop you further down the line."

This train ran every day and traversed the entire length of the Dunure & Maidens Light Railway from Alloway Junction to Girvan but returned by the main G&SW line via Maybole. It was supplemented during the potato season by a return working in the afternoon from Ayr to Maidens. The work was handled almost entirely by CR 2F and 3F 0-6-0's – the 2P on the first day being the only one I saw.

On the Tuesday morning, I was there again. The engine was a 2F 'Jumbo' and I went as far as Maidens which meant a long walk back along the track. On the Maidens side of Dunure the line passed through a deep rock cutting, single track on a sharp bend. While walking through this rather eerie canyon, the sound of a locomotive working hard became audible – and rapidly more and more audible. I fled. Back the way I'd come, three sleepers at a time, until I found a niche in the rock and crouched, terrified, as the potato special passed, very slowly but very close and very noisily. Next day I got smart and took my bicycle on the tender top.

This morning footplate ride became a daily routine during those rather unreal summers when authority was far too busy with war transport to worry about an obscure branch.

I soon learned to drive and was handling a Jumbo with the same enthusiasm as most youths show for their first car, (and three years before the legal age for that). Since the withdrawal of passenger services in the mid-'30's, there was no public transport in the area. True there was a Western bus to Dunure but it terminated there and the Maidens bus went via Maybole. So the daily freight carried passengers – in the brake van. At least usually in the brake van but sometimes on the loco as well. On one memorable occasion when shunting in Glenside (Culzean) station yard, there were the fireman, the guard, a platelayer, myself and a woman with a baby in arms on the Jumbo's footplate, (the driver was in the van). Due to my carelessness, the loco started priming and we all crowded under the engine's tiny cab roof out of the 'rain'. I'll never forget seeing that baby's head covered with black spots from the dirty water and its mother shouting "look at ma wean*".

We took water at Knowside where the Croy Bay road crossed the line. The water tank was so placed that the train blocked the level crossing for 20 minutes. The yard here, like most on the line, had a run-round loop *inside* the yard. It took three wagons and only had room for an engine between the inner points and the buffer stops. Marshalling the train was like solving one of those old *Meccano Magazine* shunting problems.

Between Knowside and Glenside there was a single facing siding immediately beyond another level crossing. This crossing was approached on a falling grade over the rather spectacular steel Rancleuch viaduct and a cutting on a curve. The crossing was protected by a fixed distant signal in each direction and the gates opened by the resident crossing-keeper – or his wife. During the summer, empty wagons would be left here and

* child

**Dunure & Maidens
Light Railway in 1945**

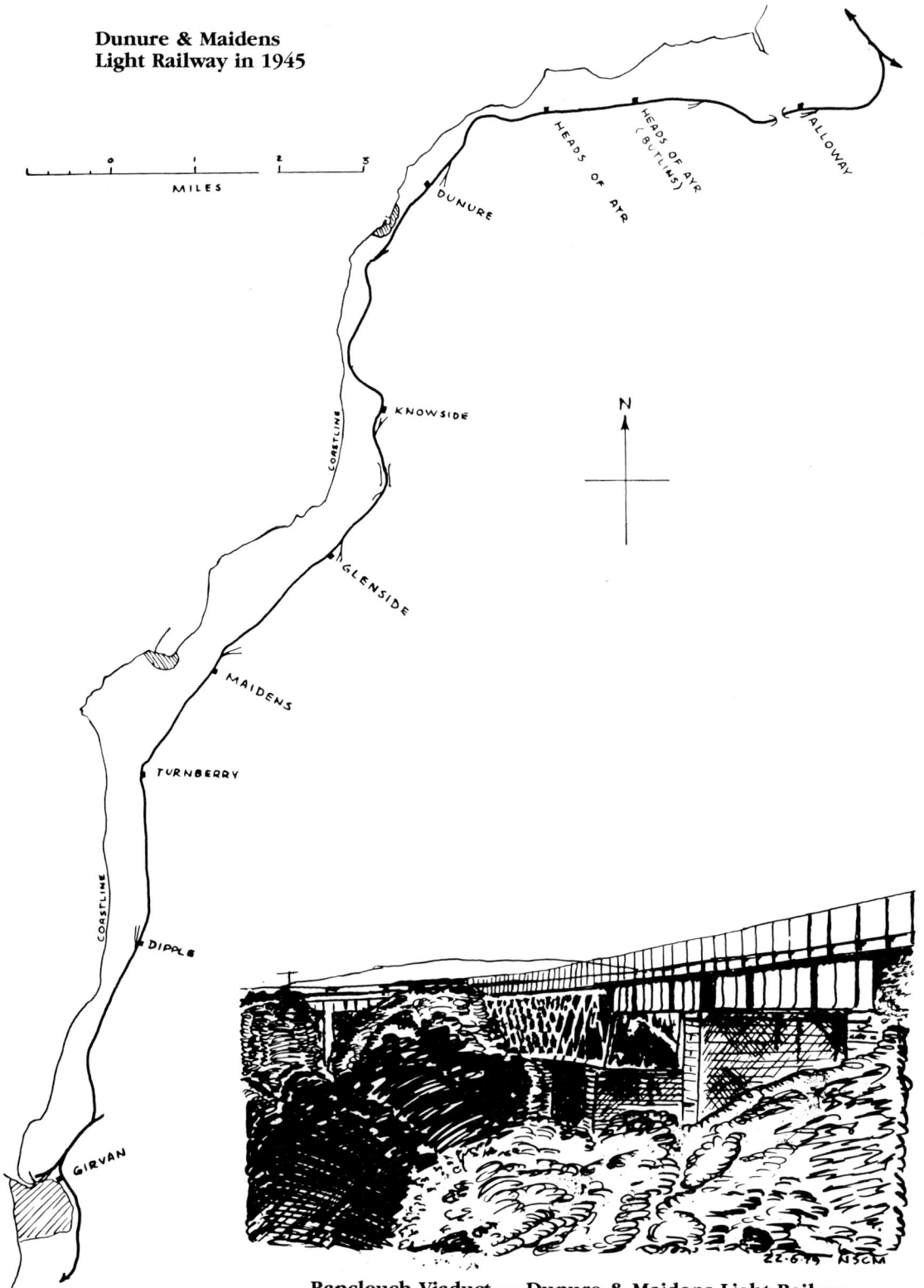

MILES

0   1   2   3

HEADS OF AYR

HEADS OF AYR
(BUTLINS)

ALLOWAY

N

DUNURE

KNOWSIDE

COASTLINE

GLENSIDE

MAIDENS

TURNBERRY

COASTLINE

DIPPLE

GIRVAN

Rancleuch Viaduct — Dunure & Maidens Light Railway

collected in the afternoon by the returning potato train. This meant running round the empties at Knowside, three at a time, and propelling them in front of the loco for about a mile with the rest of the train and van behind the tender. As the viaduct was the bottom of the dip, you had to approach this siding at quite a lick or you would stall before you got all the empties inside. Of course it had to happen! One Wednesday, market day, the gates were not open because there was no-one there to hear our whistle. There were six empties in front of the smokebox when we came skating round the bend and the gates were SHUT. There was only myself and the fireman on the footplate and he had the regulator shut and the reverse lever over and steam on again in a flash while I wound on the tender brake. Common practice on colliery pugs, it's the only time I've ever seen this done on a main line engine. The six wagons in front jerked out taut and the rest buffered up behind us as we slithered towards those gates and *stopped.* It was eight shaken people who climbed down out the brake van to see what had happened. Apparently they had been playing cards and had all finished in a heap at the front of the vehicle – so much for brake vans.

This railway was highly scenic and abounded in switchback grades and sharp curves alternately in deep cuts and high banks. An interesting feature was near the famous Croy Brae where the optical effect on the public road gives the impression of going up when in fact one is going down. The same applied on the railway and it was uncanny to drift 'uphill' with the steam off.

The yards on the line were all laid with a light rounded 2-spike chair. I suspect these might have been used on the running lines as well but by that time, conventional 4-spike CR or G&SW chairs and in some cases, LMS 3-screw chairs were in use.

The only passenger station not derelict was Turnberry. This was actually part of the Hotel and until 1942, a passenger service via Girvan had been maintained. There it was, clean and swept with its pre-war posters and no trains for the last two or three years.

There were two major steel bridges, the viaduct afore-mentioned and a girder bridge near Dunure station which, remarkably, is still there. Occupation under-bridges consisted of four light iron girders, two each carrying a timber baulk to which chairs were fastened. The chairs were of the type used in loco-shed pits and had square bases with the spikes close in to the rail. Walking over these bridges was quite daunting as the decking was usually missing.

Nothing larger than a 3F 0-6-0 was allowed until the line was cut back to Butlins' Heads of Ayr station and suddenly even class 5's appeared on passenger specials. The Butlins' trains, latter DMU's, were the line's last fling and they were eventually replaced by road transport. Near the end, an unlikely working was the LMS 4-6-2 DUCHESS OF SUTHERLAND and an LBSC Terrier 0-6-0T which went dead on their own wheels to the old Heads of Ayr goods yard, before being taken by road into the camp for display there. Even they have now gone to better homes – by road.

9. Rancleuch Viaduct between Knowside and Glenside on the Dunure and Maidens Light Railway. 26 September 1949

(G. M. Robin)

10. Ex CR 0-6-0, 57633, takes water at Knowside station on the Dunure and Maidens Light Railway, 26 September 1949.
(G. H. Robin)

11. Daily freight on the Rancleuch Viaduct. The train was stopped here for the guard to pose for his picture. This typifies the informal operation of the Dunure and Maidens Light Railway. Engine No. 57633, 26 September 1949.   (G. H. Robin)

16

Chapter 4
THE FIELD WIDENS

Inspired by my adventures on the Dunure line, I managed to befriend some drivers on local lines when I returned home. One in particular drove a CR 4-6-2T whose number I never recorded. This locomotive was by now elderly and suffering from war-time arrears of maintenance. Nevertheless it coped with its work much the same as an old horse does – no great expenditure of energy but the job quietly done. For their size, these engines had a very small cab. It was narrower than the footplate and tanks and, like most Caledonian engines, it had a large chest-high reversing lever on the left hand side. In forward gear this presented no problem, but in the reverse position the lever almost completely blocked the cab door. A visitor in the cab had to keep well out the way of the working crew.

On the Lanarkshire and Ayrshire line, passenger services terminated at Uplawmoor station but all trains ran on to a yard at Lugton situated between the L&A and the Glasgow Barrhead & Kilmarnock line. The former crossed the latter by bridge here but there was a level connection about one mile on the Glasgow side of Lugton. Local trains ran into the above yard and the engines rounded their trains and took water. There was a large signal box with a woman operator and most firemen passed a pleasant half-hour there while I helped the driver take water. The double track of the old L&A onward from here was by this time rusty and unused as a new connection ten miles on at Stevenston had rendered it redundant.

It was the practise then for rush-hour trains to leave their coaches at various outlying stations overnight and the locomotives returned light to Polmadie shed. Light engines are a gregarious lot and we usually waited for another one to appear and then run home together – sometimes there were three. All it needed was a push by one and the cavalcade would run all the way to Cathcart. These last runs home to bed sometimes reached quite spectacular speeds. On the East Kilbride branch the same practice was followed and a pair of engines, almost invariably a 4-6-2T and a 0-4-4T, ran down about 8.00 p.m. These started the sash windows rattling from half a mile away. This pair were the last steam working on the branch, by that time 2-6-4T's. A vestigal trace of this working in 1984 was two empty DMU's.

On one occasion, we were running down light between Uplawmoor and Neilston in the dark when I saw approaching us a train with a large *red* headlight. It was visible for some time before I realised what it was. An unfortunate Jumbo (Lambie 0-6-0) had found the climb out from Glasgow a bit hard on it and the lower half of its smokebox was cherry red in the dark. The saturated Caledonian engine always made a fine show of sparks at night when working hard but this one must have had its smokebox full of half-burnt coal!

On another occasion, as we approached Neilston on an evening train from Glasgow, the home signal was 'on' and a uniformed gentleman in a flat cap was standing at the trackside. "Coory doon" said the driver and I crouched down behind the reverse lever while he and the fireman leant over me on the cab opening. It was Neilston's station master. A piece of masonry had fallen off the bridge and he wanted the driver to proceed with caution. The station master climbed on to the footplate but as the engine crew made no move he was forced to stand outside the cab as we crept through the bridge and into the platform. All went well until half-way through the station, the safety valves lifted and the fireman was obliged to go to the right-hand side of the cab to put on the injector to put water in the boiler. I stole a quick glance upwards and there was the S.M.'s face just above me but he never looked down. He stepped off at the end of the platform and after a few moments' talk went off back down the train.

The regular engine, the 4-6-2T, would be off about once a fortnight for boiler wash-out and similar medications required by steam engines. On these occasions we would have another engine, sometimes another 4-6-2T but more often an 0-4-4T, occasionally a Fowler 2-6-4T or even a Jumbo. A Jumbo on that service would test the metal of the most ardent enthusiast. There were no turning facilities at either East Kilbride or Lugton and the return trip was tender first all the way. In bad weather it was unbelievably cold and in dry conditions, the coal dust blew off the low tender like a black sandstorm. I often muse on the iron men that must have driven trains before the days of cabs on locomotives.

Glasgow South Suburban Lines in 1945

DOUBLE
SINGLE
LIFTED

The big tank engines were superheated and had piston valves. It was therefore possible after starting away to 'notch up' for steam economy with the regulator open. In fact the term 'notching up' comes from just this type of reverse lever which had notches along the quadrant and a latch which fitted into them to lock the lever in position. If these big tanks refused to start their trains on a hill, it was even possible to pull the lever back into reverse with steam on and then throw it forward as the engine started to move backwards. This technique often got a recalcitrant Wemyss Bay tank on the move when all else had failed. Not so on an 0-4-4T or a Jumbo. They had slide valves. Releasing the latch on the reversing lever with steam on resulted in one of two things. Either the lever was immovable or worse, it would crash over with a force that could break your arm or in the case of the tank engine, crush you against the cab back. After starting in full forward gear, steam was shut off, the reversing lever moved a couple of notches and the regulator opened again. The operation was quite obvious to anyone within earshot and the resumed exhaust beat was much softer. As children, the less technically inspired used to insist that the driver was "changing gear" – it certainly sounded like it.

Another trip stands out. We were working an afternoon train to East Kilbride. I think it was the school train (15.17 ex Glasgow Central) and I'd just stayed on past Giffnock. Anyway, I had a shot of driving, and was quite proud of my ability to stop the five coaches exactly in the platform even though the gradient helped me. All went well till we arrived at East Kilbride and the fireman went and 'coupled off' to enable us to run round our train. Although East Kilbride was the passenger terminus by now, the line still carried on for about a mile to Mavor & Coulson's siding. It was just as well! I gave the engine some steam and we ambled over the crossover and I made a brake application – or tried to. Nae brakes! The locos were fitted with a combined Westinghouse (nearly always 'washing house' to the crews) brake and vacuum ejector, the former for the engine and in Caley days the train also, and the latter for vacuum fitted coaches now universally used in the area. I had forgotten to put on the air pump and had been stopping on the carriage brakes only, all the way up from town. Fortunately we had room to manoeuvre and no damage was done but a Pacific tank is a *big* engine to bring on to its train under a handbrake only.

About this time, two holidays are memorable for their railway interest. One was a scout camp at Biggar to which we went by train together with all our gear. On the way there I saw my only glimpse of an American austerity 2-8-0, probably at Motherwell and also a still red 4-4-0 compound which slid past us at Carluke. We all detrained at Carstairs for the Peebles branch train and while waiting there I became aware of a bright green tank engine shunting empty carriages. It was a Southern Stroudley 0-4-2T on loan. I shot off down the platform to look at it only to be summoned back by a stentorian scout-master who wouldn't listen to reason. So I didn't get a close look at it, but there was another one at Ayr as station pilot which I did examine at a later date. Amid all the wartime grime, these two engines were spotless and looked *so* different.

The Peebles branch train was worked by an LMS 4-4-2T. Of this I am certain as I saw it from camp every day and associated it with Hornby's 0-gauge product. Now LMS 4-4-2 tanks didn't come all that common in Scotland so it must have been either an LNWR or an LTSR one. An LNWR 2-4-2T certainly worked the Moffat branch and three LTSR engines appeared at Balornock shed in Glasgow a few years later.

On the return journey we again changed at Carstairs and joined a crowded express from the south hauled by a red 'ARGYLL & SUTHERLAND HIGHLANDER', my father's regiment in both wars. In their unrebuilt form, the Royal Scots were truly massive engines. And I didn't get a good look at the Stroudley on the way back either. Mention of the Stroudleys and their cleanliness brings to mind another unusual visitor. The Duke of Sutherland's 'DUNROBIN' 0-4-4 tank worked in Glasgow at Dalmuir during the war and was also beautifully kept.

The same year, another Scout expedition found some of us having to abandon a hike near the village of Aberfoyle. Public transport to Glasgow was by train on the LNER Aberfoyle branch. The line had been worked for a while by Sentinel steam railcars but by this time an ex-NB 0-6-0 worked both passenger and freight trains. Our troop nearly filled the one coach and twelve of us occupied one compartment, including a patrol leader with a bugle. No one quite knew whether he could perform on it or not so he was coerced into trying. He took a deep breath and produced a deafening blast. Immediately the brakes came on and the train screeched to a standstill in open country. To this day none of us knew whether it was concidence or not.

12. One of the Fairburn 2-6-4T which displaced the C. R. 4-6-2 tanks from local lines. Seen on a Glasgow/East Kilbride train leaving Giffnock, 27 August 1948. The new sleepers awaiting laying have in turn been displaced by welded rail.

13. Scene in the late 50's when electrification work had started. Stored wagons are standing on the Muirend/Clarkston link, 31 July 1960.

'Twas on the little local train that runs from Aberfoyle,
Eight scouts were listening to the strain of Davie's bugle call,
The boys all egged him on to play one hefty loud-long blast,
He did, and to his own dismay, the train stopped very fast,
The bugle disappeared from sight, where to? yet no-one knows,
And Davie looking very white sat back and blew his nose.
We all sat round and waited, for now the axe must fall,
Why had the old train stopped? 'Twas signalled red. That's all! *

The Aberfoyle train only went as far as Lennoxtown where an end on connection was made with an LNER suburban train hauled by a Gresley 2-6-2T.

Another interesting feature about the branch was its traversing, for a mile or so, the Forth & Clyde Junction Railway, a grand-sounding title for a *very* minor line between Balloch and Stirling. Long after the F&CJ closed, the Aberfoyle trains kept their bit in use.

A holiday of a totally different kind was at Arrochar on Loch Long about the same time. Again we went by train – on the LNER this time, always something of an intruder in Glasgow. A Helensburgh local took us to Craigendorran where another 4-4-2 tank, an NB this time, waited with two coaches. This was the Arrochar push-pull (Mother always called it push-me-pull-you from Dr Doolittle). It drew its train out of the bay platform on the steamer pier and then reversed on to the West Highland, pushing its two coaches bunker first. It had a stiff push through Craigendorran Upper and shrieked like a banshee on its NB whistle as it went through – so different from the Caley organ pipe. All the stations on the West Highland were then open, island platforms with chalet type buildings and beautifully kept, even to potted plants in the signal boxes. A great deal of that holiday was spent at the station but I never managed to chat up any of the train crews. I suppose then the LNER was almost a different religion. Apart from the push-pull, trains were handled by GNR type 2-6-0's of the K2 class and various NB 4-4-0's and 0-6-0's. A few years later, nearly all the passenger trains had green engines as the K2's were quickly repainted and the new B1's were all green. Even a Scott class LADY OF AVENEL got a green coat – vintage years indeed as two years on they were all black again.

Another haunt within easy bicycle range was the Nitshill Fireclay Works system. A private branch came off the GB&K line at Nitshill goods yard and passed round the rear of the fever hospital to the fireclay works of Alan Kirkwood Ltd. In doing so it crossed the Glasgow Corporation reserved tramway track on the level (this tramway line was quite something in itself being completely rural and largely on its own formation). After the tramway crossing there was a road level crossing over the A726 followed by a stone arch bridge over a river. Standard gauge wagons were brought up to the works on this line but the exciting stuff operated on the other side of the works. A rickety standard gauge line with sharp curves and steep undulations headed up towards a series of stone quarries cut into the rising ground. The track was made from an assortment of rails, chairs, and sleepers that bore no continuity to each other. At one place where the line skirted a minor road, one rail was chaired bullhead and the other 25lb flat-bottom – without even an allowance made for the difference in height! Over this rickety rail, trains of incredibly ancient design were worked by a 4-wheeled Planet diesel. The Planet was a successor to a line of engines going back for decades. The first was an inside cylinder 0-4-0ST by Hawthorn of Leith. This went for scrap in 1930 and was replaced by a secondhand Barclay 0-4-0ST (No. 920 of 1902). Another Barclay 0-4-0ST (No. 819 of 1898) joined it in 1939. Both these went for scrap at the end of 1946 and were replaced by a Planet four wheel diesel bought from a local scrap merchant (Arnott Young & Co.). This in turn was traded in against a new similar Planet (No. 279595 of 1949) which lasted until the end in 1960 when it was sold to the Kelvindale Paper works in Lanark and the railway closed.

The wagons were short four-wheelers with inside bearings and dumb buffers, really just wooden boxes on wheels. Just before the system folded up, many of these were renewed and these used secondhand main line wagon wheels with the outer axle ends burned off in some cases, and on others left protruding to catch an unwary leg.

* Hutchesons' Grammar School magazine 1944

14. A Fairburn 2-6-4T in full flight between Muirend and Williamwood.

15. Displaced from Glasgow suburban work by Fairburn 2-6-4T's around the time of Nationalisation, the "Wemyss Bay" 4-6-2T's gave several more useful years of service as Beattock bankers. One of the class is seen here at Beattock on 13 May 1950.

The track in the quarry proper was of necessity temporary and portable and was even more rickety than the fixed line to the works. As quarrying continued, things became even more complicated when the floor of the quarry broke through into abandoned coal workings. The workings must have been 150 years old at least and had been worked on the Pillar and Stall manner. That is to say, tunnels were driven through the coal seam at right angles to each other until a honeycomb of parallel tunnels resulted, leaving pillars to support the roof. For a time it was possible to scramble down the rubble that had fallen into the old galleries and explore them with a torch. By modern standards, there were very few roof supports - only one rough hewn tree at the centre of each intersection. Several holes appeared in the quarry floor as a result of this and two were right under the railway track. Still, they carried on with the rails shored up by timbers and the train taken over very gingerly. In the meantime, the quarry company seized its chance and made a bonanza from open cast coal working as it robbed the exposed coal pillars. Only for a short time however, as the railway was nearing the end of its useful life and dumper trucks were soon running over the railway's right of way with the rails and sleepers unceremoniously dragged to the side.

The part to Nitshill remained in use for longer and incorporated a very steep climb to the charging level of the kilns, up which the little Planet could just manage one loaded standard gauge wagon. The entire operation has gone now and ended, as far as rolling stock was concerned, a link with the early days before railway wagons were formalised.

16. Planet diesel above lime works with a "Bolsover" wagon, March 1956.

17. One of the Darnley quarry line's primitive stone wagons   (*J. Hume*)

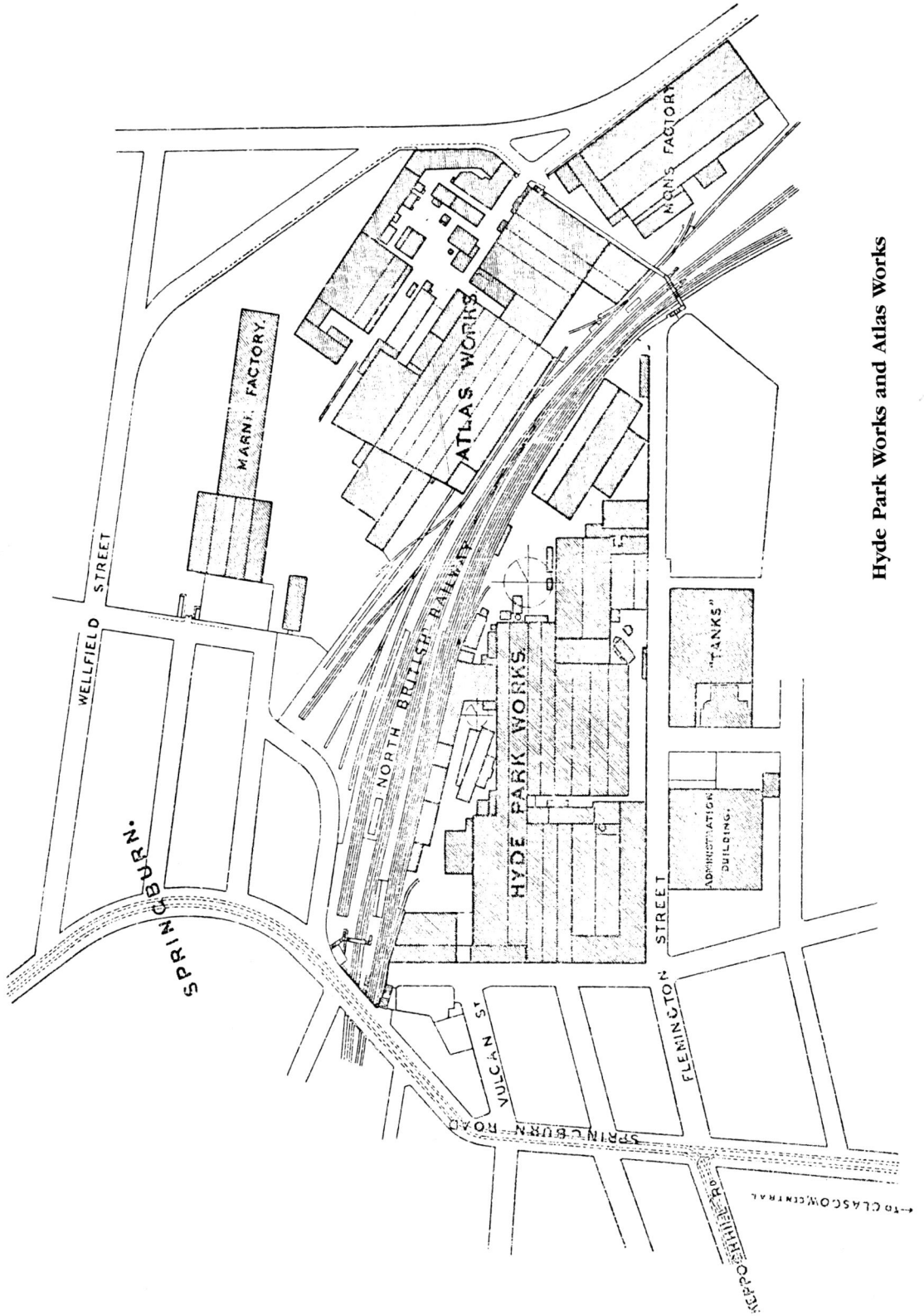

Hyde Park Works and Atlas Works

PART II

Chapter 5

DÜBSES

There was never much doubt of what I was going to do when I left school, despite my maternal parent insisting on a classical education. I became an indentured sandwich apprentice engineer at the North British Locomotive Company's Queen's Park Works. It was called 'sandwich' because the workshop bit was sandwiched between six monthly bouts of University or vice versa, whichever end one was looking from. Engineering apprentices were a new idea and had superseded pupil apprentices, the only difference being in the latter case, parents paid a fee, in the former we were paid a weekly wage like the apprentice turners, fitters, etc. I thought it was a thoroughly sporting idea. There were two of us and we were paid 29s.5d. (£1.47½) a week less 3s.6d. (17½p) health stamp, with one week's holiday at the Glasgow 'Fair'. The engineering apprentices were put through all disciplines during their five years and usually started at 18 rather than 16 as 'Highers' were a must.

The change from academic life in a grammar school to the shop floor can only be described as traumatic. The 44 hour week with Saturdays off had just started but 10½ hours seemed a long, long day. Then there was the squalor and dirt of factory life. For example one was not allowed to wash during working hours, *not* at all - not even after being at the toilet. And the toilets were almost an institution in themselves. Apart from open air urinals, most departments had to use a central block where walls were made of perforated cast iron panels. There were no doors on the cubicles and the entrance was manned by a keeper to whom one gave one's check card on entering. SEVEN MINUTES per·day was the maximum and more than that resulted in 15 minutes pay being deducted.

The only two shops exempt from patronising this central establishment were the wheel shop which was new and remote and had its own 'convenience' (still no doors but also no keeper) and the forge which was old and remote at the other end of the plant. Now the forge was one of the earliest parts of the factory and it had facilities in keeping with this status. They were situated up a stair and were also well ventilated but each cubicle (still no doors) had a bench with a hole over an iron trough which ran the entire length of the building. This flushed every three minutes or so on a time switch. Now each succeeding generation of youngsters found that if they went to the furthest 'upstream' as it were, anything that would float would be carried down the trough with the flush. And what they sent down was a paper boat which had been set alight at the critical moment. They knew full well that even a blacksmith quietly reading his *Daily Record* cannot move quickly with his trousers round his ankles. This type of establishment and therefore the story was once quite common but I think the N B had the last example.

The North British had been formed in 1903 by the amalgamation of Glasgow's three private locomotive builders. These were Neilson Reid & Co. with their Hyde Park Works, Sharp Stewart & Co with the Atlas Works just across the North British *Railway's* Springburn branch, and Dübs & Co. with their Glasgow Locomotive Works some few miles away on the 'Sooth Side' of the city. Henry Dübs had been Neilson's managing partner and had founded initially a brickworks in the Govanhill district and later started building railway locomotives there. His bricks all had a diamond motif on them surrounding his name

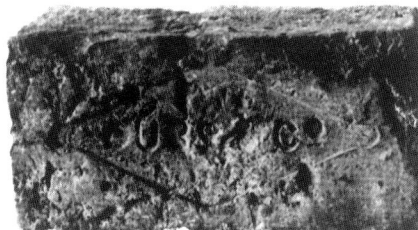

for the Railways of the World

IN almost every country of the world steam, diesel and electric locomotives built by North British in Glasgow are more than pulling their weight. The service they provide is ample evidence of the jealously guarded reputation held by us - a guarantee of even finer locomotives to be developed and built in the future.

NORTH BRITISH

NORTH BRITISH LOCOMOTIVE CO. LTD. GLASGOW

and this was used as his locomotive builders' plate. The North British kept this shape for locos built at Queen's Park until the second war and then adopted it for all products irrespective of which works produced it – it did not even require to be a locomotive.

To all Glasgow work people though, the Queen's Park Works was never the North British, nor even Dübs & Co., but DÜBSES, pronounced as two separate syllables. One either conformed with the grammatical heresy or stood the risk of being misunderstood, or worse, corrected.

When the depression hit the country in the late twenties, and it probably hit Glasgow harder than most cities, Henry Dübs' old works was closed and stood idle for a decade. It was chosen partly because it was remote from Headquarters, but more important, it had never been modernised and the original machine tools of 1864 were still being used in the main, together with some World War I veterans.

With the need for locomotives for the war effort, the works were re-opened to build Austerity 2-8-0's for the Ministry of Supply. This was not the first time the works had been called upon to help the war effort. In common with the other two works, shells and other munitions of war had been mass produced in 1914, but Queen's Park had a much more important role to play. A top secret order was placed by the War Department with Fosters Ltd., the traction engine people, for self-propelled armoured fighting vehicles mounted on caterpillar tracks. Fosters farmed this order out and the North British was heavily involved. To preserve secrecy, the order was scheduled as 'special tanks' and for the most part, were built in the Tank Shop, where in more peaceful times locomotive and tender tanks were fabricated. And the name stuck – to this day, the world over, these machines are still TANKS.

As well as the WD 2-8-0's, the Queen's Park Works again built tanks during the war. These were taken on their own tracks to a testing ground three miles away on the Cathkin Braes, (the area is now a public park). Here they were put through their paces over rough ground, complete except that they had no guns. The testing area was soon a square mile of churned earth and mud with no vegetation whatsoever. The road to the testing area through the Glasgow suburbs was quite wide as far as the village of Carmunnock, but after that it was steep and narrow. On one occasion I met an NBL tank face to face while on horseback. The poor beast took one look at this roaring monster brushing the banks on both sides and fled, me with it.

As things returned to normal, there was a tremendous amount of locomotive work allocated to these works. During my time there, orders numbering several hundreds went through – B1's, L1's and K1's for the LNER, the prototype Bo-Bo diesel for the LMS, narrow gauge RIVER class 2-8-2's for Nigeria and elsewhere, Indian broad gauge 2-8-2's of the WG class and several others. One recurrent order was for South African Mines locos. These were all the same family but each order was personalised for its company. Uusally an order was for a single loco although there was some plurality. Differences even ran to wheel arrangements – 4-8-4T, 4-8-2T and even one 2-6-2T. As the change-over to diesel power was phased in, the early orders went to Queen's Park. The first ones were 0-4-0, 0-6-0 and 0-8-0 diesel hydraulic shunters (we actually managed to sell *one* 0-8-0 to the Emu Bay Railway in Tasmania fitted for multiple unit operation! As this railway had only a handful of engines, all steam, the logic escapes me). Next came a dozen broad gauge Bo-Bo DE for Ceylon and rather more metre gauge B-B DH's for India. Following them, came the various diesel hydraulic locos for British Railways. The North British hitched their wagon to hydraulic transmissions. In many ways these are technically superior but in face of the world wide adoption of electric transmissions, they were 'on a hiding to nothing'. Making torque converters with tight tolerances on cylinder boring mills didn't help. The entire company folded in 1962 and there was suddenly a great hole in the Glasgow labour market – shortly before the end there were 7,500 on the payroll.

(Opposite) *Throughout most of its history, NBL had little need to advertise for business, much work being 'repeat' business from existing satisfied customers. The shrinking customer base of the 1950's produced a more competitive environment and advertisements appeared in the 'Trade' press; the example reproduced dates from 1955 and accurately depicts a 25 class 'Condenser' type 4-8-4 built for the South African Railways.*

NBL traditionally took the front page of RAILWAY GAZETTE's Overseas Railways special issues devoted to overseas railways generally or to railways of a particular continent. In the 1950's these featured some extraordinary inventive artwork, the ultimate being achieved in the 1952 number. Reproduced on our back cover, it depicts what looks like an early prototype of BR's eventual Class 20 traversing typically British terrain, while an Indian broad gauge WG 2-8-2 speeds past on a higher level!

18. A 1914 'works official' view of a South African Railways 12 class 4-8-2, built for heavy coal haulage on the Witbank –
Germiston line.   *(N. B. Loco)*

19. An 'official' photo of N2012, an early design of Neilson crane tank, outside the Hyde Park works. One of the type
HODBARROW No. 6 (N4004/90) survives in preservation.   *(Glasgow Library, courtesy F. Jones)*

20. The last green-liveried B1 4-6-0 ready for delivery, in early BR days. Note tallow on axle ends.

21. Two 0-6-0DH locomotives built for Mauritius Railways outside the Queens Park erecting shop in 1953. A part finished Indian railways WG 2-8-2 is on test in the background. *(Mitchell Library)*

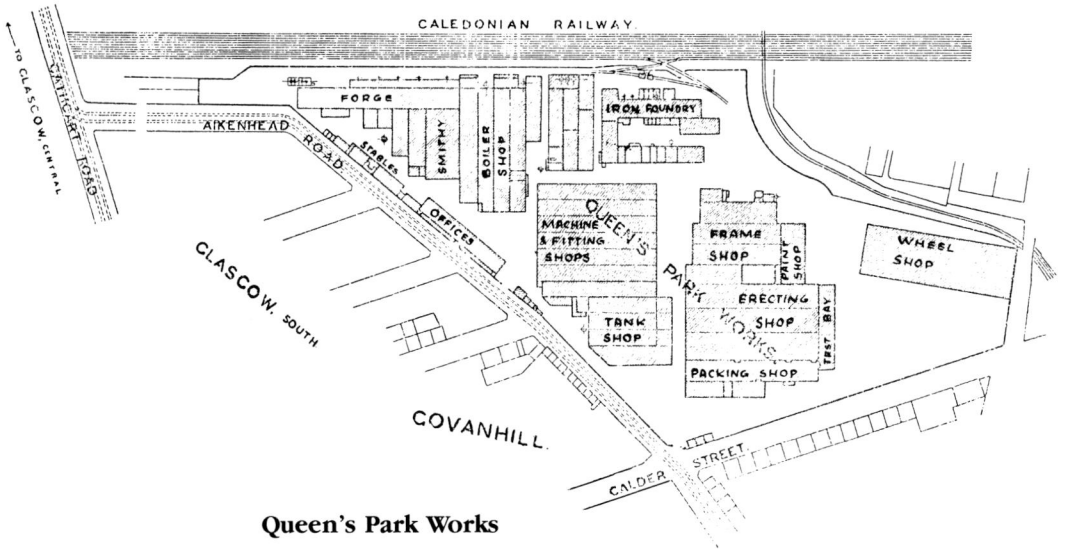

Queen's Park Works

**The Forge,
Queen's Park Works**

**Punching Machines
& Drop Hammers**

Chapter 6
HOW WE BUILT THEM, I

Forge & Smiddy

The locomotive started in the Heavy Forge. Here, marked billets of high grade steel came in and were forged into connecting rods, coupling rods, valve gear, etc., and each billet's 'birth certificate' was recorded against the finished item. No part of a steam locomotive has more operations on it than rods – each job sheet ran to pages. The billets were heated to cherry red, almost white, in huge gas ovens adjacent to the row of giant steam hammers. When withdrawn, they were supported on a central sling and manoeuvred by two or three men under the jaws of a steam hammer. The hammer driver was always a maestro; seated on a box, he had the end of a long lever in his hand. 'Up' took the hammer up and 'down' brought it down, with a neutral position in the middle. The party trick was to borrow an unsuspecting visitor's watch and bring the hammer head down on it such that he could not pull it out from under yet it was completely undamaged.

Sheet metal templates, always spelt *templets,* were used to size the billets and by the time the hammer had finished with them they were recognisable as coupling rods, etc. Several re-heats took place during the forging but for long after the operation was completed, the forgings lay glowing red on the floor and later 'hot' was chalked on them as they kept their heat for a long time. Around tea breaks (unoffical) and meal times, the latest forgings would sprout a row of tea cans boiling water. The floor in the forge was earth. No rigid floor would stand the impact of these giant hammers. Indeed, the impact could be felt not only outside the forge but in adjacent streets and houses.

The row of giant steam hammers were the most spectacular machines, but a close second was the set of drop forge hammers. On these, a die was used having a top and bottom half. Items like forked ends on rods would be made here. The top half of the die was lifted by a flat belt round a pulley, a lubricating oil splashed over the blank piece of hot metal held in the bottom half, and *wham* the top die descended twenty feet or so, stamping the shape out. The impact was tremendous and the lubricating oil flashed into fire so a great lick of flame accompanied the impact: truly an infernal machine.

Before Mr Whitworth produced his standard threads and hexagonal nuts, square headed bolts and square nuts were the order of the day (the last survivor of this practice can be seen in 'Meccano'). Now the N B didn't actually use square heads on products – but they used them nearly everywhere else as 'service' bolts. Service bolts were used to hold locomotives together prior to riveting, to clamp components to machine tables, to bolt up shipping cases and a host of domestic jobs. These bolts and nuts were produced from black round bar on some of the most Heath Robinson old machines. Basically they had a rotating turntable with holes round the periphery and were fed from an adjacent fire by the operator. Suitable lengths of bar were cut on a mechanical cropper at the side of the machine and after heating to red heat at one end were placed in the holes. As the turntable indexed round a hammer descended on the bar forming the square bolt head. Then, at the next position, a jet of cold water quenched it and it was finally ejected down a chute into a bin ready for threading. As these machines, black with age, ran, they thundered, grunted and hissed giving off clouds of steam and strained at their foundation bolts.

In addition to the spectacular machines, there were in the adjoining smiddy (blacksmith's shop) rows of lesser steam hammers, fires and anvils where a great deal of 'black' work was carried out. That is work where the parts are wrought hot either to finished items or to pre-machining shape by fire and hammer.

The foreman's office was a wooden structure above a drawing store which latter he tended to occupy rather than his office as he planned the shop's activities from the drawing prints. Apart from apprentice blacksmiths, only *engineering* apprentices had a spell of three months in the forge. The apprentice turners and fitters, although moved around, were kept pretty well to their trade. Only in the forge were any concessions granted to the next generation of engineers. As the forgemen's union would not allow apprentices from other disciplines to actually work, it was a case of keeping one's eyes and ears open and the foreman allowing the use of his empty office for writing up notes. This was great and as exams approached it was a Godsend and the *notes* tended to be more maths and thermodynamics than forge practice.

22.   On the left, two caulkers working inside a B1 boiler shell. On the right, inside a B1 firebox.

(Mitchell Library, Glasgow)

23.   Wheels within wheels. Note the stacks of tyres in the background.

(Mitchell Library, Glasgow)

Boiler Shop

Adjacent to the forge was the boiler shop, a very noisy place. The noise level from pneumatic rivet guns and caulking hammers would never be accepted by present-day factory inspectors. It was excruciating to start with but after a day or two one became acclimatised and it was even possible eventually to carry on some kind of conversation. Everybody working there suffered permanent hearing damage resulting in noise-induced deafness in later life of varying degrees, depending on experience – from ten minutes to 50 years. Old boiler makers were almost all stone deaf. One couldn't escape the noise because even at home or up on the moors, there would be a 'singing' in the ears all the time. It was peculiar, too, to be working inside a boiler shell with caulking hammers going on outside; one's own hand hammer and spanners made no sound so great was the cacophony. While I was there, there was a fatal accident when a caulker stepped back off a 6ft. high staging and smashed his head in on the floor. The police came in to take statements from witnesses. The boilermen could hear the police but they couldn't hear the men. So they took them outside to the yard to interview them and the boilermen were too deaf to make out what was being asked!

A locomotive boiler consists of a cylindrical shell with a firebox at the back end. This is literally a box fastened inside the boiler shell which is opened out to accommodate it. The box is open at the bottom where the grate is and meets the boiler at a big forged foundation ring. There is water all round the firebox and, as the sides are flat, dozens of screwed stays are used to support the sides and the top. Otherwise the steam pressure would buckle the flat plates. Between the front of the firebox and the front of the boiler, run a host of flue tubes to carry flue gases to a smokebox, added later, and thence up the chimney.

The boiler plates were once iron, but latterly steel about ¾in. thick and the tubes, once brass, were by then also steel. On boilers for the home market the fireboxes were ¾in. copper, yes – copper! The stacks of new copper plate 10ft. high must have been worth a king's ransom. And, do you know, a great deal of it went out the gate as scrap, as firebox plates are a mass of screwed holes for the stays. The boiler shells were rolled into rings, usually about three, and these would have a riveted seam and be fitted into each other like an extended telescope. Hot riveting was an art in itself. A portable fire was set up using coke and the shop compressed air line to heat the rivets (these were made on one of the Smiddy's bolt machines). A boy officiated at the fire heating the rivets. He had to heat them in rotation, giving the hottest one to the riveter. If left in too long, they burned. The red hot rivet was plucked out the fire with tongs, thrown to the riveter's assistant who caught it in the air with his tongs and stuck it in the hole. The riveter on the other side then formed the rivet with an air hammer while his assistant (called a 'hauder-on' – holder-on) held the head against the plate. So as well as noise, you had to beware of flying red hot rivets as they weren't always caught.

In 1942 King George VI and Queen Elizabeth visited Queen's Park Works to see some of our tanks being put through their paces. Following this, the Royal couple toured the plant. The Queen was particularly interested in the riveting operation and asked one of the fire boys what he was making, to which the lad answered, "Time-an'-a-hauf, Mum"!

When riveting was complete, the plates were caulked to make them tight. For this operation, a chisel-headed bit was used in an air hammer to close up any slight gaps where plates overlapped and force the edges together. The firebox stays were screwed at both ends and waisted to the root diameter of the thread in the middle. This way they could be screwed into the boiler plate and straight on into the firebox wall. These, too, were then caulked both inside and out. See where all the noise came from? An interesting detail was that the firebox stays, about 1in. diameter, had a ³⁄₁₆in. hole down the middle. This was left open at the outer end so that any cracking of the stay would show as a leak at the boiler water test.

As the boiler took shape it was laid on its side on the floor for riveting/caulking to take place. It was then picked up by the overhead crane and taken down to a clear space and rolled over on two long battens of 2ft.6in. square timber and then returned to its workplace. One of the big Indian WG boilers was having this done when the sling broke and it fell about 20 feet on to the battens. The heavy timber was annihilated by the impact but the boiler didn't seem to mind.

Fitters were employed to add various bits in the boiler shop – safety valves, inspection covers, dome covers and such. On older engines, the dome was high enough on a relatively small boiler to allow an external flange with studs and nuts to secure the dome cover. Sealing was by two concentric ⅛in. diameter copper rings in grooves which crushed together when the cover was tightened down. The studs were therefore outside the steam space. On large boilers with squat domes, this was not possible and the studs were on an inner flange so that the bottom of the stud was exposed to steam pressure. To seal it, a special fine screw thread was employed and the thread made an interference fit. It was a difficult job driving these studs but a pneumatic wrench was used to screw them in, in most cases. However, on the Indian W G 2-8-2 boiler, some nut in the drawing office specified 30 or so studs, 35 threads per inch, .010in. interference and air wrenches were *not* to be used. It's well seen he wasn't putting them in. First a tap was run down each hole, itself no mean feat. Then a staging was erected alongside the boiler and two fitters used a spanner with a 3ft. tube extension on the handle to turn the stud driver while an apprentice guided the spanner jaws round the base of the stud driver. With 1½in. thread at 35 TPI, this meant 300 or so strokes to drive one stud while the two on the spanner trudged back and forward on the staging. It took days to do one boiler.

When the boiler was substantially complete, it was moved to the erecting shop where it was finished off, the tubes fitted, and tested. It was taken from the shop on a large bogie well wagon by one of the work's pugs*. The tubes also came in bogie wagons. As one load was being lifted by a single sling, a tube in the middle started to slide out. As it came, two more started to follow it, then some more until only the rope sling was left hanging limp on the crane hook and the entire boiler mounting bay was full of giant spaghetti.

When the tubes were inserted, they were expanded into internal grooves in the tube plates and this was sufficient to withstand full steam pressure. The boilers were then hydraulically tested to twice their working pressure. Any weeps were cured by further caulking. They were then steamed to blowing off point. To be accepted, each boiler had to hold its working pressure with the regulator full open. As the boiler is just standing on the floor, this results in steam at 250 lbs per square inch emerging from a 3in. bore pipe. If the boiler shop noise didn't get your ears, this did. What's more, no-one ever said they were going to do this and if it happened when one was going about one's normal business, the effect was cataclysmic. As someone said in a much later world, "It sure makes the adrenalin flow, I can feel it runnin' down ma' leg".

* Shunting engines

24.   The "Royal Scot" passing NBL hauled by the first LMS diesels 10000 and 10001, 26 December 1949.

Foundry

Another fundamental department was the iron foundry. There had also once been a brass foundry but this was now the canteen(!) and the work transferred to Hyde Park. Steel casting was/is specialist stuff and the NB owned the Carntyne Steel Castings Co. in another part of the city.

Basically, molten iron is poured into sand moulds which have been made using wooden patterns. These patterns look like the finished casting in wood and were made in the pattern shop at Hyde Park.

Moulders are wonderful people. There must be something therapeutic about making sand castles all day and the job is peaceful and lends itself to conversation. Although it's dirty, unlike engineering dirt it washes off easily. Molasses is mixed with the sand which pervades a sweet heavy smell through the foundry. And crickets live around the sand baking ovens; big brown brutes but not the least bit loathsome. They chirp all the time, day and night. I wonder what happened to these crickets when the foundry went cold. They were supposed to be lucky.

A mould is made in a series of bottomless iron boxes later bolted together. The wooden pattern comes apart in two halves: one half is put on a steel plate, then the box over it, sand to fill and the lot rammed down. Another plate is put over the top and the whole inverted. On removing the pattern, voilà, half a mould. The other half is done the same way and the two halves when put together make a mould and the molten iron is poured in.

But the resulting casting will be solid which may be what you want, but more often a hollow casting has to be made. To do this a sand core is made. The core-makers are the élite of moulders. Cores are made in wooden core-boxes which have the shape of the hole – the negatives of the pattern almost. And making cores is like making sand-pies – very fancy sand-pies. Core sand is special with additions to make it self-supporting and it hardens when baked like well-fired bread. The core(s) are placed in the mould before it is closed and form the hole in the casting.

Big castings like cylinders had tremendously complicated moulds but the principle was the same. Only if one didn't come out right it was a big loss as each mould took days to make and fire. Towards the end of steam, steel cylinders were being cast and these were done elsewhere. The pièce-de-résistance was an order where the entire locomotive frame including the cylinders and all the support brackets were cast as one steel casting. These were ordered from America.

Iron was melted in one of two cupolas (furnaces). Each of these was a steel tower lined with firebrick. Compressed air blast was introduced at the base and the tower filled up with alternate charges of coke and pig iron together with a little limestone. After this had been cooking for some hours, liquid iron formed at the base and was led off into ladles. Smoke glass windows enabled the operator to watch what was happening. Tapping took place after lunch, the biggest ladles being filled first with small hand ladles at the end for lesser jobs where the iron composition was less critical. Ideally even the biggest mould should be filled in one go. Iron is heavy, and if poured too fast, can wash away the sand or displace the cores, ruining the casting.

At the bottom of the cupola, there was a long clay-lined spout down which the iron flowed. As it flowed, hundreds of incandescent dust particles sparked up into the air like fireworks. It's quite beautiful. To stop the flow, a clay plug on the end of a rod is rammed into the hole. Any metal left at the end of the day is run into open troughs dug in the sand floor for later use. If the iron went cold in the furnace, the entire unit would be destroyed.

The cupolas were charged from a balcony reached by an iron stair. Barrows of pig iron and coke were brought up on a hoist. This balcony had an exclusive amenity. It overlooked the Glasgow-London main line. Just after 10 each morning, I used to slip up there to watch the 10 o'clock Royal Scot and the 10.05 Birmingham go by, invariably with an LMS Pacific in front. The labourer who charged the furnace and I got quite friendly and I think I made a train spotter out of him. It was during this time that the two prototype LMS 1000 HP diesels were tried on the Royal Scot. These were numbered 10000 and 10001 only the former carrying 'LMS' but both painted black with silver coloured bogies and trim. Each morning they came by on the 'Scot' but their newness soon wore off and I eventually only went to see the Birmingham go by. It was the beginning of the end of an era.

Machine and Turning Shops

These were situated in the centre of the complex and comprised five bays of which one was the fitting shop and bordered on one side by the toolroom and tank shop. The factory had originally been laid out for all products to leave by rail. Locomotives going to the docks had therefore to leave the erecting shop on a road trailer, through the frame shop. They then did a U-turn in the yard and came back through the end of the machine shops. The passage wasn't quite straight, was slightly up-hill, and had a railway line running along it. Trailers were towed by two big ex-showman's steam traction engines owned by Road Engines & Kerr. These were in maroon livery and still had their twisted brass canopy supports. Well, these two magnificent beasts sometimes stalled when their driving wheels slipped on the rail heads. The sight of two huge engines blasting away with their big wheels slowly revolving and getting nowhere was only tempered by the showers of dust they dislodged from the roof beams, on the work, the machines and the personnel.

This rail track had another interesting feature. It was laid, as were most internal tracks in the place, with flanged dock rail. That is, the rail head is offset and a flange checkrail is rolled as part of the rail leaving a *deep* groove for the wheel flange. This rail must not be confused with tram rails which have a much smaller flangeway. In fact if you wished to run railway wheels on tram track you would have to reduce the rail gauge from 4ft. 8½in. to 4ft. 7¾in. and let the flange of the railway wheel run on the bottom of the groove*. I knew this but the contractor who relaid the floor of the two innermost bays didn't. He laid tram rail to 4ft. 8½in. gauge in 12 inches of concrete. The next time the works pug came in with two wagons she went all over the place and no-one could figure it out. And it wasn't the time to go and tell them.

To help the 1939 war work, a few new machine tools with individual motors, some of them American lease-lend items, had been installed, but when I started most machines were still driven by overhead lineshafts. The shafting was driven by an archaic electric motor which was started each morning by an electrician. Nobody else was allowed to touch it. The motor itself squatted on the floor among the foundation bolts of the steam engine that preceded it.

Each machine had a 'fast and loose' pulley. That is, one pulley which was fixed to the shafting and one which ran free on it. To start a machine the driving belt was moved sideways from the loose pulley to the fixed one by a fork operated by two cords: at least that was the intention, but the cords broke and a hammer handle did just as well. The belting from the 'fast and loose' pulleys ran horizontally overhead but the final drive was vertical by large cone pulleys to effect speed change. Pulleys were driven by 2in. flat belting and the treads were 'barrelled' to ensure the belts ran central. Belt tar was applied to prevent slip. These belts had a three-sided wire netting guard standing 5 feet high round the bottom pulley. A trick perpetuated by the wilder apprentices was to pull back the guard, grab the up-going belt with both hands and soar up into the air. At the last moment before crashing into the overhead beams or going round the pulley, they would swing over and catch the descending belt and return to floor level. The possibilities for death or maiming were enormous; one second too late at the top spelt disaster and losing grip would result in falling into moving machines or being wrapped round the bottom pulley.

One bay was the heavy bay that had the big milling machines, planes, slotting machines and shaping machines. It was significant that all the journeymen other than apprentices had a limp, or a few fingers short, or a twisted arm. Not one had escaped some kind of injury during his working life.

Planers, shaping and slotting machines cut in straight lines in one direction only. A fast return stroke is preferable and nowadays easily arranged. But in lineshaft machinery this is not so easy. The planers had two different driving pulleys, one with crossed belts to achieve reverse drive and a clutch drive to change from one pulley to the other. The shaping machines had a connecting rod driven by an eccentric wheel, but the most ingenious was on the vertical slotting machines. The gears driving their heads were elliptical with a spindle at one end so meshed that the centre distance was constant. I never quite believed them.

The lathes in the turning shop were mostly conventional, old, but conventional. Among the exceptions were half a dozen capstan lathes that had variable cone pulleys with a broad Vee belt. The distance between the sides of the pulleys could be varied. As one pulley narrowed the other widened causing the belt to run on a

* Glasgow tramways did just that!

different diameter and giving an infinite variation in speed. DAF cars used a similar thing. Also in the turning shop were a batch of multi-spindle automatics which had arrived for war work and had their own motors. These produced the special fitted bolts, stays, etc., which locomotives used in large quantities, from 'bright' round or hex bar. (Fitted bolts had plain round heads because once fitted, they were much too tight to move.) These were spectacular machines to watch and I remember being surprised to find the operators in charge of these complicated machines were semi-skilled, my first realisation that there was a rigid class structure within the factory precinct.

The machine and turning shops were lonely places. Each man was on his own and on bonus. The bonus system was basically that of a time worked out for doing every job and the man was paid extra for the time he saved. The time allowed an average of 10% bonus to be made. In theory a good man could make more but a slacker would drop to his basic rate and Mr Average made 10%. But the rate had to be worked out by someone and the ratefixers were hated. Still everybody worked like beavers. My first brush with the system occurred when I worked out a more attractive way of turning some special washers. I made a fixture out of scrap metal that allowed me to do 30 at a time instead of singly – in more or less the same time. I finished off the entire 50,000 in a couple of days and clocked off the job to the tune of 150% bonus feeling great. Now this was apparently a good paying job, around 20% and I had finished the lot for the order, so no one else got a chance at it. Then the ratefixers said "Well done sonny" and adopted my method at a new time that would just make 10%. That was my first lesson in how *not* to win friends and influence people.

The turning shop also worked nightshift month about, other departments tended to have constant nighshift. The factory was a different place at night. Shift started at 9.30pm and worked till 7.45am. Nightshift workers were allowed to clock out at 7.40 and make for the main gate to allow the day shift workers a clear way to clock on. When the whistle sounded, the gates rolled back and the nightshift swarmed out into Aitkenhead Road like a dam bursting. The boiler shop handcart went that way one dark morning with four lads kneeling on it while another pushed so as there would be no hole showing in the sea of heads.

Light Turning Shop

Hydraulic Press

Lathes & Sharpeners

**Corner of
Machine Dept.**

  I learned one thing on night shift. When using a hand file on a lathe, you file left handed. I tried filing in the normal way and my left sleeve caught in the chuck. I was lucky. I found myself lying in the passage with no sleeve - it could have been no arm.

  The apprentices in the machine shop formed a squad within the shop under an old soldier called Alfie who was supposed to keep an eye on us. He looked after the tools for the apprentices' machines and these were kept in a wooden cupboard. The inside of these doors had various bonus times chalked up but one large chalk number puzzled me. Every day it had been rubbed out and was re-chalked one less - and on Mondays three less. To start with my ignorance was a great joke but eventually one stalwart confided "It's the number 'o days to RA FERR" (the Glasgow Fair) which was the works annual holiday, one week in those days. Excitement grew as it got down to 10 - - - 3, 2, 1 and off. A week later the number was up again - days to New Year.

  Among the apprentices were a number of mixed Indian and western blood. Britain was pulling out of India and these people ran the railways there, and had done so from their inception, a colonial solution to an embarrassing problem. Their culture was European and they referred to Britain as home although they had never lived here and suddenly they were going to be alien in their own country. They were all bright lads and I made good friends with some but they all drifted away later as our careers developed.

Fitting Shop

This was next to the heavy machine shop, in fact the drilling machines spilt over into it. Sub assemblies were put together here but the main job was finishing coupling and connecting rods and valve gear. By this time these had been through their umpteen machining operations but the sharp edges had all to be radiused off by hand filing and emery cloth. They were sent off to the erecting shop as if they had been satin chrome plated. Each item carried the engine's running number stamped on it. They went out on the finished locomotive the same way. It used to make me weep to see them later on, rusty and dirty, after the blood, sweat and tears we'd wasted on them.

Wheel Shop

A kind of outpost of empire, this was situated in a far corner of the yard, divorced from other buildings. A wheel set consists of two wheel centres, in our case usually spoked, two tyres, two crank pins if they are drivers, and an axle. Some may also have roller-bearing axleboxes in which case they go with the set. The centres are machined on lathes by normal turning methods and are pressed on to their axles by hydraulic pressure. Once on, they should never move again. The same applies to crank pins. However, some designs call for a fitted key as well. The keyway should be machined in the wheel and axle to line up but they didn't always and

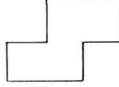

somebody, usually me, got the job of taking a 2in. square key 6 inches long and filing it to this shape so that it would fit flush and not be visible.

Once the wheel centres were on the axles, the tyres would be fitted. These were consumable and after 3in. or so wear, during which time they would be periodically re-machined, new ones would be fitted. The tyres were placed inside big gas rings and heated to a dull red. The pair of wheels was then up-ended by the overhead crane and one wheel dropped into the hot tyre. It was then allowed to cool and contract on to the wheel. When both had been attended to, the complete wheel set was set up in a wheel lathe and both tyres machined to profile simultaneously. A wheel lathe had two headstocks and a gap between them big enough to take the largest driving wheels – in our case 7ft. 0in. diameter.

Driving wheels were then set up on a balancing machine and lead poured into pockets in the crescent-shaped balance weights.

A new thing about this time was roller-bearing axleboxes. Both Hoffman with their taper roller bearings and Skefco (SKF) with their spherical roller bearings were trying to get in on the act. Both types were highly sophisticated pieces and somewhat out of place in that environment. Skefco's probably had the edge as they were self-aligning but Skefco wouldn't trust shop floor labour to fit them. They sent their own man who came in the staff entrance in an Anthony Eden hat and donned a white boiler suit. However, he always got an apprentice to help him, in this case, me. The roller races themselves came in fours, wrapped in greaseproof paper and packed in wooden boxes. They were cleaned in white spirit with clean hands and heated in oil before sliding them on to the previously prepared journal. New sheets of white paper were spread on the floor under them and they were hand packed with grease. If a dollop of grease fell on the paper it was left there lest it was contaminated. After that the split housing was bolted to the outer race with the same care.

The empty wooden boxes were piled up as high as the crane rails. During a quiet spell, I realised I was alone trying to look busy, when the old labourer who ministered to us told me to "go away in with the others"; he would watch for the foreman. This pile of boxes was hollow! By taking out two next to the wall, an opening led into a den in the centre with boxes and sacks for seats where the rest of the squad were playing cards or sleeping. It was ideal for swotting for exams. Then one day, Skefco's lorry turned up to take them all away.

Frame Shop

Adjacent to the erecting shop was the frame shop. Frames were flamecut on an automatic machine that had a magnetic follower on a template. After that they were bolted together in batches of two to ten depending on thickness and the edges machined all round on a multiple head slotting machine, one operator to each head. The frames were then paired off and sent into the erecting shop. At least usually, but sometimes the frame shop got ahead of itself and piles of frames were stacked out in the yard. This happened to the frames for the LNER/ BRK1 2-6-0's. A hundred and forty frames were stacked in the yard and the weight pushed the bottom ones into the ground. Shortly after No. 69's frames had been taken in, a consignment of plate was offloaded from a railway wagon on top of the last two K1 frames. They were never found and a new set was rushed through. Some years later I was down from the drawing office and saw them exposed and unwanted. They may be there yet.

25.   South African Mines 4-8-2T in grey livery. Note absence of paint or rear coupling which would not show in photograph, January 1952.

26.   Another version of the South African Mines locos, this time a 4-8-4T for the East Rand Proprietory Mines. November 1951. The white painted patch on the paint shop wall is to avoid touching up the "daylight" under the boiler on the official photograph.

Chapter 7

HOW WE BUILT THEM, II

The Erecting Shop

The Erecting Shop was by far the most interesting part of the works. Here all the nondescript pieces of hardware came together and emerged, often under its own power, as the finished locomotive. Unlike most railway company works where the tracks run along the length of the shop, the N B built their locos across the three work bays. Two of the bays were for engines and the third was for tenders. The tender bay also housed the exiles from the boiler mounting shop already mentioned and was host to the LMS 800 HP diesel as an almost permanent resident. Also of interest in the tender bay was the Sentinel steam engine for the company's steam lorry, recently displaced by several Foden eight-wheel lorries.

Another obvious difference from a company shop was the absence of tracks over the pits, except for the four roads that ran, two each, to the paint shop and test tracks, as only by that time would an engine have wheels and no repair jobs were contemplated. The pits with tracks had cast iron multigauge rail reflecting the company's versatility in rail gauges and the others, plain iron slabs. The rest of the floor was wooden setts rather than the concrete used in other departments. At the finished end the multigauge track continued into the paint shop but no further, and into the yard for the length of the paint shop where they converged. Only the standard gauge rails went through the points here but at least one broad gauge diesel tried it. The foreman's office stood for all the world like a signal box, between the boiler-mounting section of the tender bay and the middle bay. During the period in question the senior foreman (of three) was one *Mr* Bulloch, instantly shortened to 'the Bull' when out of earshot. Under his office was a materials store where various consumables were kept and issued. Conspicuous among these were candles. Only the motion squad had an electric inspection lamp which they guarded jealously. Everybody else used candles. They were issued in 4in. lengths lest anyone got carried away with extravagance and the art of making candles from the stubs of old ones was soon mastered. All the work underneath engines or inside tenders and sidetanks was carried out by candlelight. Tanks have baffle plates with manholes in them to prevent water surging, so it wasn't funny to knock the candle over if you were deep inside a tender. Especially as the manholes were not in line for the same reason.

In the centre of the shop between bays 1 and 2 was a blacksmith's fire. The smith's job was to adjust the length of brake rods and such like as no two locomotives turned out identical. It is fairly obvious that he could lengthen a rod by heating the middle and hammering it on the anvil. But he could also *shorten* one as well. If asked to take 2⅞in. off a 6ft. long rod with forked ends at each end, he just went ahead the same way. And you didn't have to measure it. It was 2⅞in. shorter, straight and true, and unless it had been painted you couldn't see a fire or hammer mark on it. In the winter there were always a few 'having a heat' at the fire until a whispered cry went up, "Here's the Bull" and there would be a mad scatter. Another dodge was to heat your spanners on cold mornings, and if the big doors were open for an engine being tested, it could be very cold. But you had to watch any tools you left near the fire or someone would make sure they were heated and they would be too 'heavy' to lift!

The very last pit in the middle bay had, at the finished end, a set of rollers on which a completed locomotive would be tested. The coupling rods would be removed and the driving wheels allowed to rest on the rollers. The engine could then be run while any adjustments to the valve events could be made. This was a most impressive operation. A large smoke duct was supposed to take the exhaust away.

Locomotive frames came from the Frame Shop and were 'squared' at the starting end of the two lines. This entailed bolting the frames to their stretchers or cross members with service bolts, our own smiddy ones, until a more permanent fastening was made. Squaring was carried out by using a large L-square through the axlebox guides which had already been fitted in the frame shop.

Locomotive frames divide roughly into two quite distinct philosophies. The plate frame, used invariably on all British engines and most places influenced by the British Raj, dates back to Stephenson's time.

Erecting Shop   **Cylinder Shop**

27. Indian Government Railway broad gauge 2-8-2 No. 8301 in photographer's grey. Note multigauge test track, 28 May 1950.

Initially made of wood with sheet metal facings, it soon developed into two vertical flat sheets of some depth and varying in thickness about 1in. to 1½in. depending on the size of the machine. The frames are deep enough to overcome the weakness introduced by large square cut-outs for the axleboxes. These cut-outs are re-inforced by horseshoe-shaped axlebox guide castings and fitted for stays or keeper plates, but the main design stress is through the frame itself. This type of frame has the cylinders mounted either between the plates where they form a stretcher, or riveted to the flat of the outside, and in a few cases both. This type of frame also carries the locomotive's running plate and cab.*

These frames came from the frame shop with all the necessary holes jig drilled out undersize, say ⅞in. for a finished 1½in. hole and ¾in. service bolts would be used in half the holes. After squaring and service bolting, the ⅞in. holes would be opened out to their finished size by drilling and reaming in situ. Two techniques were used but both used a combined taper drill and parallel seamer. In the hard way, a driller and his mate had a big heavy portable air drill which one of them could just about lift. It had a big extension bracket to hook over a suitable piece of the frame and a jacking screw to force the drill forward into the hole. A 5 foot long tube was used to counter the drilling torque. The driller's mate hung on to this for dear life lest the drill took charge and maimed the pair of them. In use these drills snarled away but did a surprisingly good job. Incidentally all portable drills and grinders were pneumatic as electric ones were deemed a security risk. The more civilised approach was the use of a large horizontal drilling machine with a heavy base and column-mounted drill head. This was on wheels and, when positioned beside a frame, could be jacked up and do several dozen holes before moving on.

Cylinders were fastened by service bolts and a sheet metal 'finger' with a saw slot in the end was bolted to one of the cylinder head bolts. A piece of string with a knot was wedged in the slot and calipers used to ensure it was in the centre of the cylinder bore. The string was then taken through the bore and tied tightly round a large square clamped to the driving axlebox guides. Calipers were then used to check the concentricity of the string to the bore at the *rear* cover end. If it did not line up, the cylinder was given a whacking great 'dunt' with a 14lb hammer until it did. Once truly aligned, the holes were reamed out.

The frame and cylinder assembly were riveted together with fitted *cold* rivets. These were a force fit in the holes and were fitted from the inside using an air jack. The outside was then riveted over into a countersink by a pneumatic hammer (more noise).

* except in BR standard designs.

28. An apprentice working on the "rod squad". Coupling and connecting rods had more individual operations than any other components. *(N B Loco)*

Why the North British was building so many *bar* frame locos for the colonies at this time is not clear. Perhaps where people could compare the two types, it was proving superior. It certainly, by the middle of the 20th century, was a better 'engineering' job but much more expensive. Always considered the American type of frame where it was used exclusively, it originated with Bury, a rival of Stephenson, in the days when the London & Birmingham Railway was being built. Bury supplied dozens of little 0-4-0 and 2-2-0 tender engines at this time and their frames were quite literally two bars of iron to which everything else was hung. Bury's engines were so small that within a few years of the opening of the L&B several engines were employed to pull each train. This limitation, and Bury's refusal to consider enlarged designs, together with the greater prestige of Stephenson, caused new construciton of Bury's engine to taper off and with it, the bar frame technology.

29.   Locomotives for the African colonies and dependencies were a significant part of NBL's business. In South Africa, British-trained engineers and draughtsmen developed distinctive indigenous designs to cope with the rugged conditions found throughout the Cape, but, lacking locomotive building facilities in the country, contracted construction out to private firms in Britain, Germany and North America. NBL, building on an association formed by Neilson and Dübs as far back as 1880, captured a large share. In the post-War years construction for SAR included 19D 4-8-2's, S1 0-8-0's, 24 class 2-8-4's and GMA/M Garratts built under sub-contract from Beyer-Peacock. These apart, the most impressive design was the 25 class 4-8-4, some of which were built with condensing apparatus for use in waterless areas of the great Karoo. NBL were proud of the 'condensers' and featured them in contemporary advertisements (see pp 26 and 77). This 1953 'works official' view depicts a conventional 25 which incorporates sundry American-influenced design features such as cast steel underframe incorporating the cylinders, mechanical lubrication and self-adjusting wedges.
(N.B. Loco)

30.   The 15F class were a piston-valve version of the earlier 15E mixed traffic 4-8-2's, and became the most numerous SAR class, eventually totalling 255 engines. Most were equipped with mechanical stokers. NBL delivered their first 15F's in 1939 and despite wartime pressures produced a further batch in 1944. The final examples, SAR 3057-31546 (NBL29541-26040) were delivered in 1948.
(N.B. Loco)

However, some of Bury's engines were sold in the United States where their small size again told against them but the infant American locomotive industry copied Bury's construction technique and adopted the bar frame. Again it was literally a bar, but as the classic American 4-4-0 developed and held sway for 50 years, the frame evolved as two parallel bars each side over the length of the coupled wheels symmetrically above and below the axle centre line with the axlebox guides acting as spacers. From the leading driving axlebox a single bar each side extended forward to carry the outside cylinders and pilot beam. The cylinder castings incorporated half the smokebox saddle and were bolted together to form a complete cylinder/smokebox saddle which straddled the frame bars. The boiler and smokebox were used to stiffen the entire frame assembly and a 4-wheeled bogie kept the front end up. This arrangement, together with a flexible springing system, proved ideal for pioneer work on the lightly laid rail* of the period. A contemporary British 2-4-0 or 'single' would have cracked its frames in no time.

The bar frame as finally developed was flamecut from 5in. or 6in. plate but had little vertical height. The driving axle centre line was on the frame centre line and the top and bottom bars were formed by cutting away large areas of plate between the axles. Forward of the leading axle the frame continued as a solid section profiled to clear the bogie wheels and to carry the cylinder block. A similar extension rearwards supported the firebox and rear drag beam and was profiled to clear the trailing truck. Cylinders straddled the frame and supported the smokebox and, as the frame could not possibly do so, the boiler also carried the running plate and cab. The motion brackets for the valve gear were carried by two cantilever beams across the *top* of the frame between adjacent driving wheels, allowing the valve gear to be supported more rigidly than usually possible with a plate frame.

Bar frames were not riveted – they were bolted together. Service bolts were used as described but reaming was by the portable drilling machine. It would be a stout man that would hand drill 1½in. holes through 9 inch metal. About 200 precision bolts were used. On the Nigerian 2-8-2's they were parallel and .010in. oversize. On the Indian WG 2-8-2's they had a slight taper as had the hole, but were still oversize. All the bolts were driven by hand-swung 14lb hammers. The bolt heads were plain round heads with a considerable chamfer. To prevent them being damaged by the hammer, a large block of metal supported by a thick wire handle (⅜in. dia. wire!) was interposed between the hammer and the bolt head and held by a 'hauder-on'. One mark on the bolt head and the inspector would condemn the bolt and it would have to be drawn out and nobody wanted that – but nobody! It took 30 to 40 strokes with the 14 pounder to drive one bolt home. When a frame was ready for bolting each man in the erecting shop would come and drive one or two bolts. For the one squad to drive all the bolts themselves would have nearly killed them. After each bolt was driven, the hammer wielder was invited to whistle – few succeeded. A normal right-hand swing was straightforward and remarkably satisfying. But some bolts were in awkward situations and had to be swung from above, between one's feet, or from a kneeling position above one's head, or left-handed. There were one or two 'corry fisters' or left-handed gentlemen in the shop and their services were heavily in demand for hammering some bolts. Once home the bolts had washers and castellated nuts with split pins fitted, and finally, three hefty 'darts' across the protruding threads with a chisel. The finished frame complete with cylinders looked truly massive and made the adjacent classical British frames look like tin cans. American locomotives always had a reputation of a hard life but a short one and it is difficult to reconcile this with their frames. Especially as some British locos were phenomenally long-lived despite cracked frames.

Each engine was built by an erecting squad, usually a fitter and an apprentice. There was also a motion squad, a wheeling squad (on permanent nightshift), a smokebox squad and a cab squad. This may sound as if there was not much left for the original pair but in fact these specialist gangs only spent a short time on each engine and the bulk of the work was carried out and 'progressed' by the erecting fitters. It took about two to three weeks from the frame stage to steaming and about two engines per week were turned out.

Each bay of the erecting shop had two 30-ton overhead cranes on the lower crane rails and a 90-ton crane on the upper rails for lifting complete engines. The relative shortage of lifting power meant the services of the crane could only be obtained when the job was too heavy for two men to lift. One particularly nasty job was the ashpan assembly on the LNER K1 2-6-0's. Although looking very similar to the modern B1 4-6-0 also being

* In many cases a mere 25lb per yard.

31. A finished Indian W. G. 2-8-2 is carried down the erecting shop for loading on to a road trailer. On the right a similar locomotive is merely frames and boiler.

(N.B. Loco)

32. All hell let loose! A B1's boiler in steam in the test bay. The regulator has just been opened and the main steam pipe is open to atmosphere.

(Mitchell Library, Glasgow)

built concurrently, they had evolved from a much earlier design by rebuilding. In fact some castings had G N R (Great Northern Railway) on them which we were obliged to grind off. The ashpan could not be fitted until the boiler was on the frames as, unlike the B1, it would not pass between the frame plates. It took four men to lift it and manoeuvre it under this frame which would be standing on jacks. It was then lifted amid oaths and groans until it entered the space under the firebox. An apprentice then wriggled up through the ashpan door from the pit and put four service bolts in and the rest released. If the supporting quartette dropped the assembly the apprentice was trapped and probably hurt but the longer he took the more voluble the oaths. As an older apprentice due to staying on at school longer, I was often treated as a journeyman and did both the holding jobs or went inside the ashpan as required. I was truly glad on that score to see the last K1 out the door.

Another awkward job was positioning cast-iron frame stretchers (the B1's had fabricated mild steel ones) between the frames. These would be lowered on a rope sling and a fitter and the boy, one on each side of the frame, would catch it through the holes with a tapered iron bar called a 'podger'. This was used to pull the holes in line and then service bolts fitted. Then the crane would slacken and the sling would be removed. On the first K1 frame next to us some difficulty was being experienced in lining up the holes and the fitter told the apprentice to hold his podger in the hole while the fitter explored the hole with his finger. The boy misunderstood and put *his* finger in a hole on his side and removed his podger. The heavy stretcher sprung ever so slightly on the sling but it was enough and they were both looking aghast at the stumps of their index fingers, neatly guillotined between the stretcher and the frame. The apprentice found his in the pit and carried it around for a time in a matchbox until it started to smell.

We had a man-sized hexagonal nut problem. This was long before metrication, unified threads and American threads added their problems. As mentioned earlier, service nuts and bolts had square heads, but Whitworths' standard pre-war had a larger hex size for British Standard Whitworth (BSW) nuts than British Standard Fine (BSF) of the same shank diameter. Spanners had their sizes stamped on them, eg. 1½" BSW, 1¼" BSF. This was fine except that during the war, all hex sizes were dropped one shank size to save metal so the above spanner would fit 1¼" BSW or 1⅛" BSF. But some post-war orders, namely Indian, specified pre-war hex sizes when after the war BSW and BSF were both fixed as the same hex size. Add to the nut confusion that spanners were marked with the sizes pertaining to the time they were made and you never knew what size to look for in the tool locker. "Go and get me a ⅝" spanner" could take the afternoon to resolve. I still have spanners in my car tool kit from those days that only fit my (metric) car by accident and perpetuate the old pre-war legend.

As a locomotive developed, it was lifted stage by stage down the bay, gradually acquiring its boiler, cab, wheels, valve gear and all the fittings that go to make completion. One of the last major operations was the assembly and setting of the valve gear. This was carried out by a rather elite squad who kept their secrets to themselves but whose job was no more complicated than that facing any model engineer. The locomotive was stood with its driving wheels on drums which could be motored round and the travel of the piston valves (no slide valve engines were built by the NB during this period) was measured and adjusted.

As mentioned above the locomotive was first steamed on the drums but then had its coupling rods put on and was run up and down one of the multigauge test tracks. These tracks were only 100 yards long and rather inadequate. They both ended in a curve which checked bogie clearances, but one particular order for Nigerian type 2-8-2's had a curved length of metre gauge track using secondhand BR chaired bullhead rail laid across the three bays. Engines were not completely finished when they were tested. For example, they were seldom painted. Or rather they were a piebald mass of brown undercoat, grey filler and black, as the painters worked incessantly in the erecting shop painting, filling and rubbing down everything they could see. Engines seldom were tested with their own tenders. On the drums they either had none or an old coal wagon, but on running test the same tender would be used for several successive engines. If an LNER B1 tender was there, a Nigerian 2-8-2 could be seen coupled to it chuffing back and forward on different guages of rail. That would be a coal-burning 2-8-2. Some were oil-fired with a large, shaped tank fitted into the tender coal space. To test and set the burners, one of the tanks was mounted in an old coal wagon. The oil burners had a special grate and no ashpan. On one occasion a burner was extinguished and oil poured through the firebox grate into the pit and immediately caught fire there. Flames enveloped the entire back end of the loco and started on the wooden coal

33. A Nyasaland Railways 2-8-2 charges out of the erecting shop on trial.

34. South African Railways design of 4-8-2 supplied as an industrial order for Transvaal Navigation Collieries, January 1952 (compare with illustration 18).

wagon and no-one could get on to the footplate to turn off the fuel. Fortunately before things got too much out of hand, the burner re-lighted and the spilt oil burnt out. The burning wagon was doused before its load of fuel oil exploded and continued to be used albeit a bit charred looking.

One of our own squad's first engines that I worked on was K1 No. 62002 and we had been running it back and forward all morning. When everyone went for lunch, I was last on the footplate and duly screwed down the tender handbrake before leaving it – good mainline practice. After lunch, I was detailed to another job and the K1 dragged its tender back and forward with its wheels locked while everyone scratched their heads as to why it was so sluggish. Sometimes old nuts and things got left inside steam passages and would be ejected violently through the chimney during their trials. In fact one K1 came back in shortly after delivery with a smashed cylinder. A 1¼in. nut had found its way into the cylinder and was found mangled between the broken piston and the hole in the cylinder head.

Orders for South Africa were always 'strippers'. Durban did not have any cranes big enough to lift a complete engine so the boilers, frames, sidetanks and wheels went out separately. The Nigerian type 2-8-2 which was being supplied to various East and West African railways went complete but with couplers and cowcatchers removed and bolted inside the tender. When the last unfortunate K1 came to be wheeled, (remember its frames were lost and replaced), its driving wheels were missing only to be found in the packing shop with wood packing round the journals and crank pins and stencilled 'L965 – DURBAN'. Goodness knows what South Africa would have made of three pairs of 4ft. 8½in. gauge wheels – all South Africa is 3ft. 6in. gauge or less.

The last loco of any batch was always a nightmare. Fittings in the small parts stores were drawn out as required but if something did not fit, the stores bin would be rifled until a 'good' part was found. By the time the last loco was being assembled all the good parts had gone. It was not too bad if the badly fitting pieces had been chucked back in the bin. This merely meant filing holes oval or grinding off corners, but more likely the offending part was lying in a pit somewhere where it had been thrown in disgust. There was nothing for it but to make the offending parts by hand, to wait for new ones to go through the system would have held the job up indefinitely. No additional time was allowed for the last engine so the cheapest labour was used – the apprentice. Great ingenuity was shown in making up brackets, pieces of footplate steps, handrails and so on from scrap. But I think a set of wheels would have beaten even the best apprentice.

With the 100 WG 2-8-2 engines for India the loss of small parts was wondrously solved, for the erecting shop anyway, by an order for 100 boilers and parts to make a further 100 locos in India's new workshops, an early DIY kit. Loco orders had an L prefix, tenders T, and all spares orders D. D8570 was a number soon engraved on the hearts of all; for the erecting squad because it was a gold mine of extra parts, and for the spares people because it was a running sore. Virtually no part of D8570 made it to the packing shop intact. This meant raising a new spares order for the missing/stolen/scrapped/lost parts and even this had to be guarded all the way through the shops. A batch of 100 items would easily lose 2 at each of six manufacturing operations. Even in the packing shop pieces would mysteriously vanish. When it all, or nearly all, was crated and shipped, management heaved a sigh of relief. The packing shop worked 36 hours non-stop while the spares manager stood guard. They finished at 6 a.m. and decided amongst themselves to go home, leaving the crane driver to clock them all off at 7.45. This unfortunate fell asleep in his cab and only awoke when the day shift came in at 8 so there was another row. And I have a feeling that somewhere those handling the NB loco's erstwhile affairs will still have an open file numbered D8570.

The same applied to a lesser extent on the South African Mines locos. If parts didn't fit too well on a 100+ order they would all be re-machined as a batch, but the Mines' engines were nearly all 'one-off' orders, so if something did not fit or had not been made when erection commenced, it had to be done the hard way with chisels and files. Using a hammer and chisel is a lost art. Once my left hand had healed up a bit I got quite good at it. You don't hit your hand with the hammer after the first day or so and a sharp chisel can cut cast iron as cleanly as a machine tool. Years later I had the pleasure of showing a young journeyman toolmaker in a sophisticated modern factory how to use a chisel. I did the first of a dozen cuts to show him, dressed in my drawing office white coat, while he looked on incredulously. He finished the job but afterwards I noticed with a smile his black and blue knuckles.

35. South African Mines 2-6-2T loco. One of the Author's efforts, 10 December 1949.

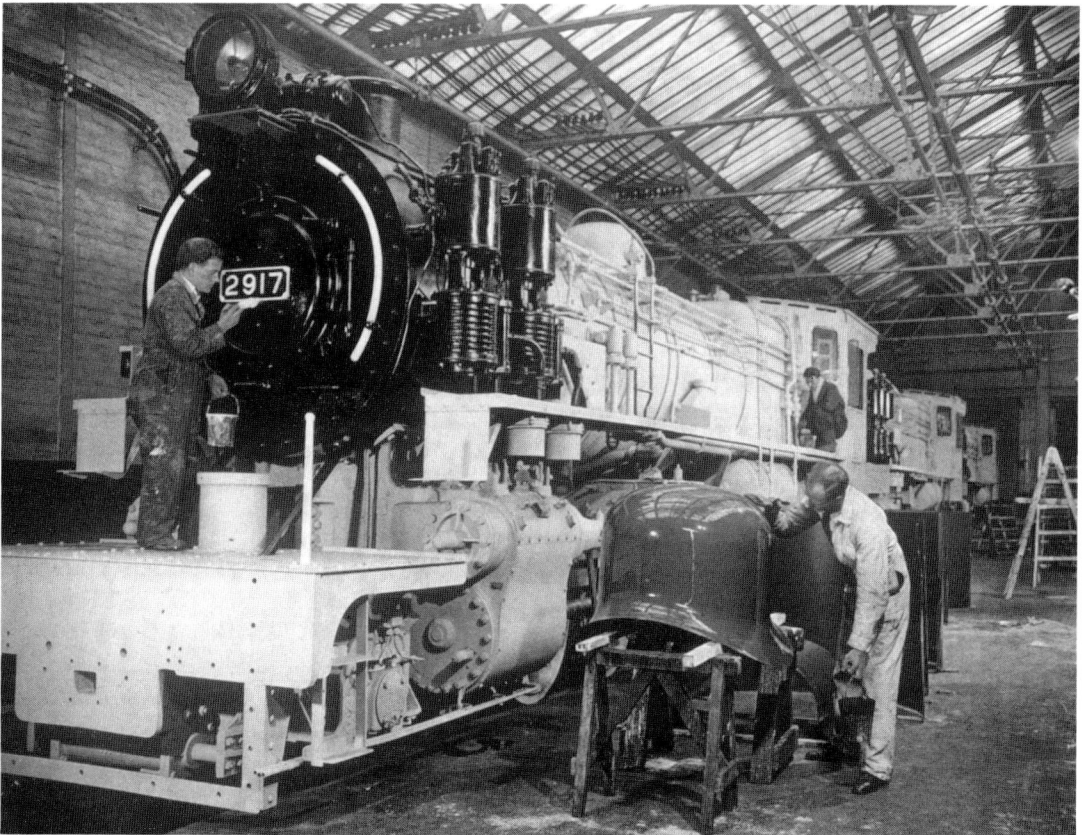

36. EAR&H 2-8-2 in paint shop. Note the absence of coupler and cowcatcher. The dome and boiler clothing is for the 2-6-2T shown in photo 35.

(Mitchell Library)

The South African Mines locos were odd things in many ways. One order was in fact a pure South African *Railways* design of 4-8-2 *tender* loco of a type built before the first war. Although large engines, especially when one considers any industrial locos bigger than an 0-6-0T unusual in the U.K., they were primitive. Sometimes they were not even super-heated and their styling belonged to an era decades earlier. In most cases they were 4-8-4T or 4-8-2T with no flanges on the *leading* driving wheels. Cabs, bunkers and tanks varied from one to the other.

One pleasant variation was L965, a saturated 2-6-2T for Simmer & Jack Mines Ltd. There seemed to be no particular hurry for this little engine and one of the squad fitters and I were allocated its erection as a fill-in job when other work was held up. This in practice meant that the fitter had other things to do (backing horses, football pool collection, etc.) while I got on with putting the beast together. During this period the charge hand wanted a model yacht made as a lamp stand and I spent a happy, if somewhat confined, afternoon sitting up inside the casting of L965's smokebox saddle shaping and sanding this yacht hull. One odd thing about this engine, it had roller bearing crank pins quite out of keeping with the Mines loco concept.

As these engines were all strippers they were never completely finished and usually steamed without the cabs complete, no couplers, etc. Their parts were finish painted after dismantling unless an official photograph was required. Then they were painted light grey, lined and lettered in black and white on one side only and posed for the photographer. Parts for the picture would be laid in place and bolt heads on the cab sides would be just that – no fittings on the other side. Sometimes the same Mines locos would be photographed with several different letterings to avoid duplication of effort. Even a K1 received this treatment but later a finished black picture was taken of another one.

The Paint Shop

The paint shop led off the erecting shop through big wooden doors. These were usually kept shut as the paint shop was well heated and quiet. Small finishing jobs were carried out here, tightening pipe clips, putting in split pins and the like. In winter it was nice to be warm and remarkably peaceful. Painting was a lengthy process. First a coat or two of grey undercoat on top of the filler, then several coats of colour. After the colour, lining and lettering was carried out using dead matt spirit paint from a palette. To get lettering and the curves of lining correct a card stencil with pin holes along the lines was used. It was held against the loco or tender side and a chalk 'pounce' patted against it. The pounce was a muslin bag full of chalk and enough of the chalk came out of the bag to leave the outline of the stencil on the paint. Next, straight lines were generated by holding a chalk-covered piece of string against the panel and 'twanging' it. A small brush with 3in. long hair was then used for the lining and a palette brush for letters and numbers. Transfers were only used when specified, e.g. BR lion totems on later BR engines. After lining, several coats of varnish sealed the spirit paint and completed the 'coach finish'. On a completed engine you could see your face in the paintwork.

After tightening pipe clips along the left hand running plate of a black B1, I inadvertently marked the cab side with the toe of my boot. Nobody noticed so I whipped out a piece of cotton waste (incredibly hairy stuff) and tried to polish out the mark, only to find the varnish was WET. I just fled and escaped. The entire cab side, lining, number and all, had to be taken down to the undercoat and re-done.

Orders that were stripped were painted piecemeal so there would usually be a pair of side tanks, a bunker and other parts all finished, painted, lined and lettered. When it came to the boiler, this was different. What the public sees on a finished engine is in fact only the thin steel clothing which covers the thermal insulation and it is held on by the boiler bands which fasten underneath. Strippers' clothing therefore when painted looked like huge open ended green drums. The boiler bands, which were usually lined out were painted on the flat. Many painters of model locomotives do this and feel slightly shamefaced, but there's a prototype for everything.

37. The ancient Dübs crane tank in QP works yard. A tender tank for a Nigerian 2-8-2 is on a flat wagon to the right of the engine, May 1949.

38. The NBL built Neilson 0-4-0ST (NBL 16351) outside the paint shop at Queen's Park Works. *(Collection – F. Jones)*

39.   The Andrew Barclay 0-4-0 crane tank (AB 955/02) at the Atlas Works c1952   *(Collection – F. Jones)*

40.   Neilson Reid 0-4-0ST shunts wagons past old stables into forge, 13 March 1952.

41.   Unpainted LNER B1 No. 61363 on test in yard of the NBL. 6 March 1950.

42.   A CR Jumbo 0-6-0 stands disconsolately in the NBL yard after causing a trial of destruction through the paint shop, 6 March 1950.

The Yard

The yard was out of bounds to all workshop personnel unless there was a good reason for being there and the yard foreman made it his personal job to enforce this. However, at this time, all foremen wore a brown coat and a hat, so he was easy to spot. I had one spell of three weeks in the yard. An order for Spain took the form of castings, plates and materials only and someone who understood drawings was required to identify all the stacks of Imperial sized plate, decide what metric parts they had been ordered for, and have them marked. It involved shifting tons of material so there was a squad of labourers there as well as the works' only mobile crane. It was midwinter and bitterly cold so there was a coal brazier made from an old oil drum. We were all standing round it getting a heat when I realised the shipping clerk's trousers were on fire, really blazing, and he was still talking to me. Only then did I find out he had an artificial leg acquired after a wartime plane crash.

There was a complete standard-gauge railway system in the yard. All heavy pieces were moved from shop to shop on a variety of railway wagons. For large pieces like boilers, there were two heavy bogie flats. Lesser pieces were moved on four-wheeled flats with dumb buffers and NBL diamond maker's plates on the solebar. There were also a few dozen mineral wagons of the typical Scottish type with cupboard doors and single brake blocks. Unlike the flats, these were painted red oxide and were labelled 'NBL'. At one time they may have plied on the mainline but by this time they just lay around the yard or transferred coke from the stock pile to the smiddy. To shift these wagons were two early products of the NB. The elder, built by Dübs in 1893 No. 3080, was an 0-4-0 crane tank which wheezed around and even used its crane on occasion. It was helped by a Neilson standard 0-4-0ST built in NBL days in 1904 No. 16351.

These Neilson saddle tanks were that company's stock line for many years and superseded their older 'box' tank. They appeared all over Scotland and some were extremely long lived and have eventually been preserved. In slightly modified form, principally a longer wheelbase, they appeared on both the Caledonian Railway and the North British Railway, one of the latter still existing.

Then over one New Year holiday (1951) the Dübs tank disappeared and was replaced by another older Neilson 0-4-0ST, No. 5934 of 1901. These engines were usually both in steam all day and were forever wandering through shops with a wagon or two to the general annoyance of the inmates. The rest of the time they played with the coal wagons in the yard. Wagons being moved from one shop to an adjacent one were often propelled by an overhead travelling crane using its cross carriage with the hook under the buffer beam. This provided enough impetus for the wagon to roll into the next shop.

Every morning, there was a 'shunt' from the main line. At one time there had been two gates giving access, but latterly only one. An old Caledonian 'Jumbo' 0-6-0 would propel its train back through our gate, round a 90° curve on steep gradient and into the reception sidings. It would hang around a bit, maybe even shunting, and would leave with any empty wagons that were ready.

Now when a British locomotive was ready for delivery, it was moved out of the paint shop to raise steam where it wouldn't soot up other paint jobs. It then sat outside the closed doors on the track leading into the paint shop. One morning, a B1 was ready and the points from the paint shop had been set but delivery was postponed until the afternoon. At 10.30 the Jumbo propelled its fifteen or so wagons into the yard and ran them into the B1. The Jumbo was still out on BR tracks and the driver could not see this. But he felt the extra weight and assumed this was merely the curve and the gradient taking effect. With "it's heavy today" to his fireman, he opened the regulator and the old engine, slipping like mad, pushed the B1 back through the *closed* doors into the paint shop. By the time the B1 had travelled the length of the space it had recently vacated, it was ambling along nicely. The next loco was an Indian WG which it pushed into a similar one behind it, crumpling up the overhanging cab roof on its smokebox. This second WG's cab cut a neat hole in the *closed* door into the erecting shop where a B1 standing ON JACKS brought the whole cavalcade to a stop, fortunately without falling over. The LMS driver and fireman eventually arrived, one walking up each side of their train, to stare in awe at the devastation. The driver was incredulous that his small 50-year old machine could pick up 300 tons of locomotive on top of its own train and get it all moving. The only casualty, strangely enough, was a painter who had been working on top of the second WG and realised the cab was not cutting a hole for *him* through the erecting shop door and jumped, spraining his ankle.

43. The NBL's pride, the demonstration 0-4-0 diesel hydraulic shunter trapped behind the derailed Barclay crane tank. Mixed gauge track is for testing underground mine locos, 2 April 1952.

44. NB Loco diesel hydraulic shunter in service at a Glasgow steelworks, 31 July 1961.

**Track Layout,
Queen's Park Works**

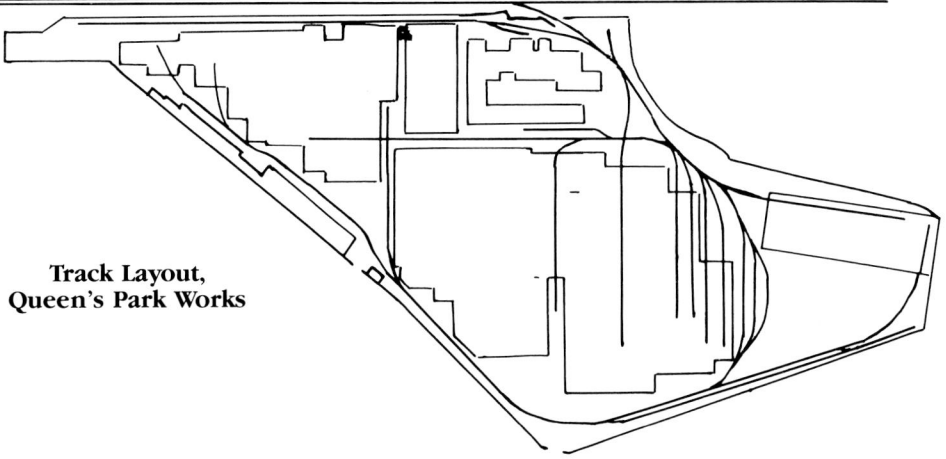

45.   Indian State Railway 4-6-2 class YP being loaded aboard ship by the giant hammerhead crane at Stobcross quay. The building
in the background is one of Glasgow's Universities.   (*Mitchell Library*)

46. A new B1 is seen off by the faithful. The yard foreman has just appeared and we are all *outside* the gate, April 1950.

47. 61369 rests at Helensburgh on its trial run before returning flat out to Eastfield, May 1950.

The track in the yard was old and extremely rickety but it was all relaid and where applicable re-paved in 1950, about twelve years before it was abandoned altogether.

Hyde Park Works did not have a track *system*, only a few sidings off the LNER Springburn branch. A solitary Peckett 0-4-0ST No. 1477 of 1917 sufficed here. It spent a brief spell at Queen's Park in 1955.

On the other side of the line, the Atlas Works had a great deal of trackage and employed a quaint 0-4-0 Crane Tank built new for the works by Andrew Barclay Sons & Co. of Kilmarnock in 1902, No. 955. The NBL often passed their smaller orders over to Barclay who were a much smaller concern and built in ones and twos, where the NBL built in hundreds.* The other Atlas pug was a spanking new NBL 0-4-0 diesel hydraulic shunter No. 27078 of 1950, painted a bright green and lined out. It was used for demonstration purposes when potential customers appeared. On one such day, the Barclay crane tank de-railed itself trapping the diesel on a siding. The visitors could only look at it static. We did *not* get that order. However, it was a very powerful little loco for 200 HP and had better starting characteristics than contemporary diesel electric locos.

Trials and Testing

Locomotives of other than standard gauge could not be tested except on the short lengths of multigauge track in each works. Engines for the British market were despatched on their own wheels in steam via the LMS tracks behind the works. The engines had an LNER (later BR) driver and fireman plus an NBL fitter and sometimes an apprentice. I made it my business to be that apprentice as often as possible. The route from the Works to the LNER shed at Eastfield was over a network of freight only lines and took nearly all day. These lines went through Glasgow's East End which then had some pretty rough areas. We were pelted with ballast once while waiting at a signal and the driver told me of a time one hard winter when he was working a coal train in the area. The local wild men poured oil on the rails as he approached, causing the engine to slip to a standstill on the gradient. While the crew were struggling to keep their train moving, the Apaches climbed on to the wagons and dropped the wagon doors, allowing the coal to spill out on the trackside. By the time the 'polis' arrived there was neither coal nor soul to be seen but Blackhill was warm for the rest of the winter. Now, BR don't even try to run trains on that line. Arrival at Eastfield was always a thrill as the new locomotive in its shining paint contrasted vividly with the dirty blacks all round. Even a well-kept Gresley Pacific looked tawdry beside a sparkling new green B1.

The following day, the engine was run down to Helensburgh on the former North British Railway line. Normal service to Helensburgh was worked by 3-cylinder Gresley 2-6-2T engines on rakes of four Gresley twin articulated suburban coaches. The West Highland trains used the same line as far as Craigendoran where they swung away to the right. There was plenty to see.

The new engine was very carefully oiled and prepared and driving on the outward journey was done with frequent stops for checks. Tallow was pressed in the Vee centre in the wheels and put atop the tender axleboxes. It could be clearly seen as a white spot in the wheel centre but if a bearing ran hot it would melt and disappear. At Helensburgh, a final check was made and then HOME – like a bat out of Hell! With no train and a clear road, a B1 can compete quite favourably with a Maserati, at least for thrills. There were always several people on the footplate and one sunny day I sat up on the shelf behind the water scoop to keep out of the way. From there I could comfortably enjoy the ride and look out over the eave of the cab roof. At Cardross tunnel I crouched down until we were through but a mile later on, I felt a 'pressure' near my head. This was the serrated edge of Dumbarton Station awning passing my scalp at 80mph.

Occasionally, foreign locomotives ran trials on home metals. One instance that stands out were two diesel hydraulic 0-6-0's for Mauritius. These ran on various duties for one week in 1953. Tests of starting heavy coal trains were made on steep gradients both as single units and together in multiple. For the last two days, they ran on passenger trains between Glasgow Central and Edinburgh Princes Street. For 0-6-0's with coupling rods and small wheels, they ran beautifully steady at 75mph. Similar trials were carried out with the LMS prototype 800 HP diesel electric painted in Works grey. By that time BR had happened and it was numbered 10800.

---

* The NBL is no more, but for a time, Barclays' notepaper carried 'incorporating the North British Locomotive Co.' in small print on its letter-head. Barclays are now part of the Hunslet Group.

48. LMS 800hp diesel loco leaving NBL Queen's Park works for initial trials, 5 June 1950.

49. Two Mauritius Railways 0-6-0DH locos in Glasgow Central Station on a Glasgow–Edinburgh test train, 11 November 1953.
(Mitchell Library)

Chapter 8

POLMADIE

The N B L canteen was pretty basic, but the canteen at the L M S Polmadie motive power depot was only too pleased to feed any of Dübs personnel that would come. An increased turnover meant a more varied menu and for the period it was an excellent canteen. It even provided cutlery. It could be entered from the street or the running shed so what could be more natural than for a young fitter in a well used boiler suit, having entered by the street door, to leave by the shed door along with everyone else and escape for ten minutes into the magic of the running shed.

Polmadie Shed was particularly interesting at this time. British Railways had just been formed and the new order was just taking effect. One development was the shedding of our L1 2-6-4 tanks ordered by the L N E R in this, an L M S shed, for running in. When the K1 and later B1 orders were built they reverted to Eastfield as the L M S drivers preferred their own engines and objected strongly to the L N E design with its long narrow firebox. Then the new liveries were just appearing. First, a few engines appeared after general repair with 'BRITISH RAILWAYS' painted in white and the letter 'M' painted above the number. Very soon after this the new B R numbers, i.e. the L M S number with 40,000 added, became general, starting, at Polmadie, with a Fowler 2-6-4T. At the same time the large class '3' number on L M S carriage doors was dropped; the L N E R had anticipated this by some years.

The first real sensation was DUCHESS OF BUCCLEUCH turning up in cobalt blue with LNWR red, cream and grey lining, shortly followed by a few more of the same class. A Jubilee class 5xP also appeared from Leeds painted L N E R apple green but with LNW lining and it looked very smart.

New Fairburn 2-6-4T's were being delivered at this time. They were one of the few unlined classes to receive the Ivatt LMS livery. No. 2272 was the last to have standard LMS yellow serif letters with red shading. Nos. 2273 to 2276 had Ivatt straw letters and numbers with a maroon edging and 2277 had this style of number but no letters. From then on it was 'British Railways' but very soon the new locos were being delivered fully lined out in the black LNWR livery which was later to become so common. Gradually local engines would get the LNWR livery and, as the first of each type received it, it was fascinating to see how the application improved them. Engines like the Caley 0-4-4 tanks and 4-4-0's had been black since 1928 and unlined since the outbreak of the war. The new style seemed to suit them well. One class of old friend departed, though. The entire remaining set of Wemyss Bay 4-6-2 tanks was banished to Beattock for banking duties where they ended their days. The new Fairburn 2-6-4T's not only ousted the 'big pugs' but also displaced the Stanier and Fowler 2-6-4 tanks. The latter were not missed. As one wit remarked, "the 4-6-2 tanks were foul machines but the 2-6-4's were even *Fowler.*"

Part of this daily visit to the shed resulted in striking up friendships with some of the drivers and, this time, *they* offered *me* footplate rides. While most Polmadie engines were not well kept, some of the 2-6-4's working the south side suburban services were kept immaculate by their drivers. Copper pipes were scoured and polished including the large balance pipes between the bunker tank and the side tanks which showed under the cab. Smokebox straps and buffer heads were painted white or silver, never, I'm afraid, scoured and polished, but nevertheless eye-catching. One of the drivers who had 2243 arranged for me to make a large brass star for his smokebox door so the N B L erecting shop came to a standstill while Nigel made his star. He gave it back to me when he retired and I have it yet. These decorations were a tradition that went back to the early days of the Caledonian. This particular driver had no time for nationalisation and when his engine came back re-numbered from St. Rollox Works, he and his fireman got cracking with scouring powder and turpentine and soon had the offending white letters and numbers rubbed off and the shaded LMS 2243 showing beautifully. Only the 42243 on the smokebox number plate showed the new order. As she had officially been re-numbered, it was only years later that a general service resulted in a complete paint job sinking the LMS livery for good. I'd some good runs down to Wemyss Bay at weekends on this engine. Wearing a worn boiler suit, the 'fitter' on the footplate was never challenged.

50.   Fairburn 2-6-4T No. 42243 at Wemyss Bay station. Note "British Railways" and new number rubbed off and immaculate condition of the engine. Author and driver Tommy Farquarson on running plate, 1 February 1950.

51.   Dixon's Ironworks Ltd. Govanhill, Glasgow, 0-4-0ST No. 4 (AB 689/91) on the branch from Polmadie tip, 21 March 1956.   (W.A.C. Smith)

After a while, BR standard classes started to arrive at Polmadie and the big engine liveries settled down, first to bright blue and then to Brunswick green. The BR locos to cause the biggest stir at Polmadie were the 'Clan' class 4-6-2's. They had *chime* whistles which the LMS had never heard before and they didn't settle down for months until everyone in the shed tried them and I think all their friends and relations as well. Many of the NBL overseas products had chime whistles, but people living near Polmadie shed must have been heartily sick of them before they settled down. It would seem impossible that, right at the end of the steam engines' career, a bad locomotive could be designed but these Clans seem, by all accounts, to have been a disaster. Still they looked nice.

One practice between Glasgow Central and Polmadie was to run light engines and freight trains nose-to-tail on the freight only lines without the benefit of the absolute block signalling system. At busy times congestion built up until a continuous line of stock stood all the way back to Larkfield while trains on both the fast and slow lines in both directions pounded by. One solution was to send engines into Central coupled together and on one occasion five 2-6-4 tanks set off together, each one with a different paint style. On another occasion, as a crowd of us crossed Polmadie bridge on our way to lunch, there was an almighty bang and we looked over the parapet to see a Jumbo sitting quietly on the main crossover, not on the rails but on the sleepers. At the busiest time of the day it blocked the up and down fast lines, the up slow, and the up goods with its queue of trains. Unfortunately I had to go back to work, so I don't know how they sorted it out. And it looked so peaceful squatting there simmering quietly to itself but absolutely immovable.

Just beside Polmadie shed there was a brickworks. This used clay from an old pit, level with the Rutherglen end of the motive power depot and a narrow gauge railway ran alongside the railway fence for about a mile. Internal combustion engined locomotives were used on this and they plied up and down continuously. But in those days, narrow gauge was not yet respectable and I never photographed it or explored it fully. On one occasion however, while lunching in the shed canteen, we watched the roofs of their buildings burning merrily and apparently unnoticed.

The third major activity at Polmadie was Dixon's Blazes, a giant iron works in the nineteenth century idiom. When Dixon started in the area there had been a chain of collieries feeding into the works but by this time the last of them was standing derelict and the very extensive private railway system he employed had been largely dismantled. The LMS west coast main line between the NBL and Dixon's was one of his earlier works having been built as the Pollok & Govan Railway and later taken over by the Caledonian.

Dixon had a weird collection of industrial 0-4-0 saddle tanks, some of them reasonably presentable but some utterly disreputable. A few were built by Dixon or cannibalised from parts of others. There was also a Grant Ritchie 0-4-2 saddle tank. One line of the system, still open, took slag from the furnaces to a tip ('coup' in Scotland) in a marshy area known as Moll's Mire. This involved a bridge under the main line, a sharp curve past our works and then a half mile traverse across waste ground. A 0-4-0 saddle tank would take these ladle wagons of slag slopping over the brim, to the imminent danger of anyone happening to walk along the line, which had become by way of being a right-of-way. The ladles were tipped by a chain attached to the loco which was uncoupled and steamed smartly away to activate the tipping mechanism. Tons of hot slag spewed down the tip and at night would light up the sky for miles around. Tramps would seek the warmth of the cooling slag on winter nights and once, one of them slept too long in the morning and was entombed in molten slag.

Dixon's also had hundreds of their own mineral wagons stored on the site. It is doubtful if any were still acceptable to the main line authorities, certainly they never moved out during my time there. Several had been pushed out over the ends of sidings and straggled in uneven lines across rough ground. They were red oxide with DIⓍON painted in white, the X being opened out to encompass the side doors.

Dixon's ran its own coking plant and when a coking furnace was ready for emptying, one side was removed and an hydraulic ram pushed the entire contents of red hot coke into a very large steel hopper wagon. This was then propelled by a small diesel loco under a quenching tower where the coke was doused with water. From the top of this tower dense clouds of steam then churned upwards and the thermal impact must have been colossal. One lunch time, three of us nipped over the main line to watch this process more closely. It really was an awesome sight. This too lit up the sky at night as part of 'Dixon's Blazes', but I suspect the nickname was coined when the iron furnaces were open-topped in the early days.

CORAS IOMPAIR EIREANN
1600HP Diesel Hydraulic
Locomotive

Scale ½" = 1 foot

North British Locomotive Co. Ltd.
Glasgow

D53-20

Chapter 9

THE DRAWING OFFICE

In due course I was transferred to the drawing office in Springburn. With the forming of the North British Loco. Co. all the office functions of the three constituents were centralised in the new admin block at 110 Flemington Street. This was a four-storey building arranged round a quadrangle. The front entrance, in stone, was impressive and had the front of a locomotive carved above the door. The number was styled like a locomotive number plate. Inside was a splendid entrance hall with large models of some of the Company's products and a grand staircase going up from it. I was rather chagrined to find that only senior staff could use this entrance and stair. Lesser beings entered from Adamswell Street at the back where there was a time office, and four plain stairs, one on each corner of the building, were the normal way up or down.

The drawing office occupied the entire top floor of the building. The front section of the square and one side were the steam drawing office with private offices for the Chief Draughtsman glassed off along the front edge. The other side comprised the spares drawing office and the diesel section, while the tracing office formed the remaining side of the square and was partitioned so that we couldn't see the girls. There were about 100 draughtsmen and 50 tracers at this time so, by any standards, it was a big drawing office. Beside the tracing office, the only other girls in the D.O. were two secretaries and a tracer employed in the project office, a glassed office top secret section beside the Chief Draughtsman's room. This tracer was an extremely well-shaped young lady but as she worked behind closed doors, was out-of-bounds. A technique evolved by the younger draughtsmen was to arrange that prints for the project office were ready for collection by this Goddess just when the sun would be shining horizontally down the passage between the diesel and spares sections. As the print room was in a corner of the main tracing office, the progress of the girl down the hundred feet of office with the sun behind her cheered us all up and showed her shape to perfection.

As time went on, the diesel office grew and the steam section dwindled. The diesel moved into the tracing office and the tracers took over the side limb of steam. By moving from the shops to the D.O., I leapt forward two years in the progress in orders – no diesel contracts had percolated down to the shops as yet, apart from the experimental ones. Career wise, the diesel D.O. was a good move but the steam D.O. would have suited me better. Unfortunately I had read a paper to the Institute of Locomotive Engineers advocating Rapid Transit Systems which had impressed the Chief Draughtsman with my modern ideas.

The drawing boards were all flat boards supported on wedges on wooden benches. Beside the board was a side table formed by an extension of the bench over storage drawers. Senior men also had a back table supported on the bench behind them and sometimes used to support a second drawing board. Loose tee squares were used on these boards and we provided our own squares and instruments. Pencils and rubbers were issued by the D.O. clerk as if they were gold bullion. Pencil holders were provided for stumps and new pencils could only be obtained on production of the old dog-end which was kept lest you tried to get a new one before the one issued was completely used.

At one time drawings had been done on cartridge paper but only the General Arrangement drawing of an engine was so done and even this was discontinued for diesels. The G.A. drawings were kept in a basement drawing store and dated back to the founding of the consitutent companies. The really old drawings were works of art. All lettering was in copperplate handwriting and the various materials were depicted in different colours, round surfaces being shaded. In the early days, no detail drawings were issued and this one master drawing contained enough information to build the entire machine.

Early in the twentieth century, the practice evolved of drawing on translucent paper and using a photographic process, printing the drawing through on to light sensitive paper. This immediately opened the door to the modern practice of issuing detail drawings of every part and distributing these to all concerned.

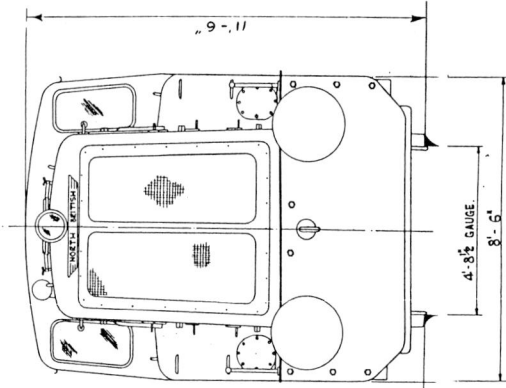

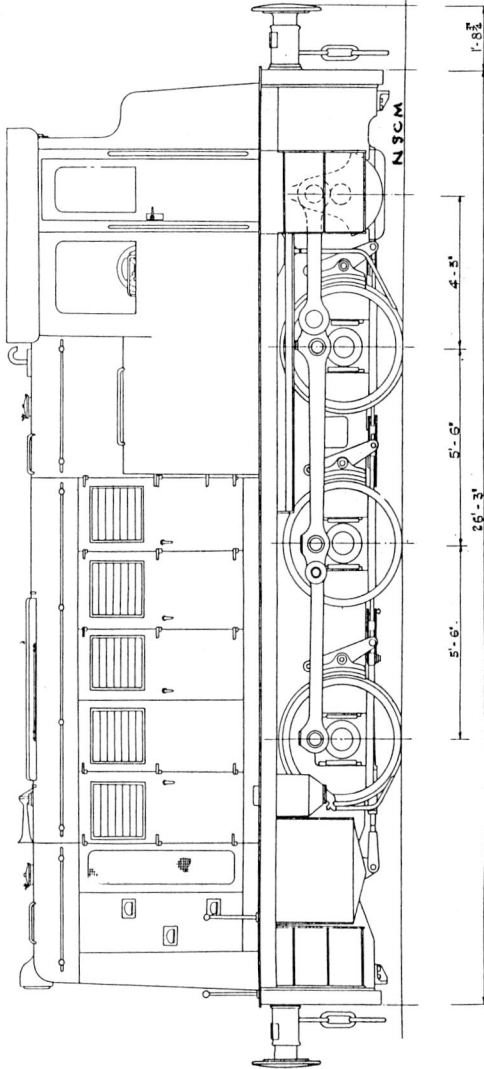

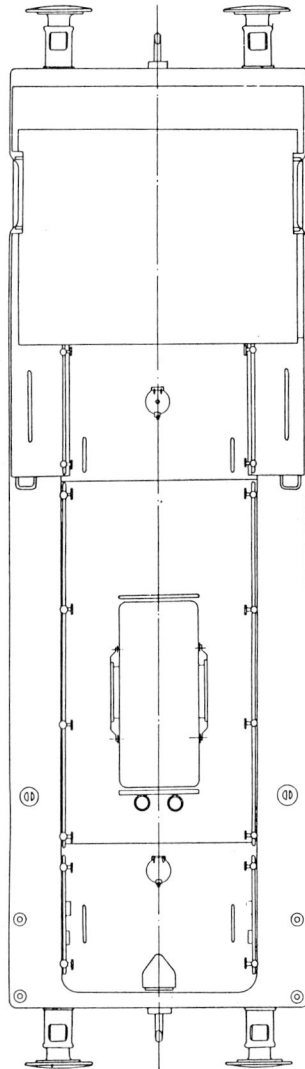

300HP D.H. Shunting Locomotive
N.B. LOCO Co.

Early printing paper gave a white line on a dark blue background, commonly known as a 'blueprint'. It will be seen therefore that a blueprint is not an original document as commonly supposed, but merely a transitory copy. Modern prints employ a dye-line process which gives a dark blue line on a white ground and is much easier to use.

Pencil on drawing paper however does not give a very good print and drawing paper deteriorates when handled or exposed to sunlight. Hence tracing – the tracers put a sheet of starched Irish linen over the pencil drawing and copied it through in Indian ink. The tracing cloth looked not unlike blue polythene sheet and superb prints were obtained from this. The girls had no idea what they were copying. They worked on a flat drawing board with no wedges. First they traced all the curved lines, then all the horizontals, then all the verticals. Any mistakes were copied through too which could be disastrous as any but small alterations damaged the surface of the cloth. The cloth tracings had to be treated with care as any creases printed through and moisture rendered them opaque. Sweaty hands could leave permanent fingerprints and a cup-ring from a tea-cup was there for ever. No draughtsman ever puts a cup down on a drawing.

In 1951, the diesel section was completely snowed under with work and unable to expand quickly enough to cope. Accordingly, some orders were put out to contract drawing offices. These were 0-4-0, 0-6-0 and 0-8-0 diesel hydraulic orders. The locomotive engineering required for a diesel or electric loco is less than for a steam one, as the engine, transmission, electrical gear, brake systems, etc., tend to draw on other engineering disciplines. This enabled contract offices not skilled in locomotive lore to forge ahead with only one NBL draughtsman keeping an eye on them. They retained the original drawings of each order until it was completed and supplied us with six prints of each drawing for issue to the shops. As the detail numbers on drawings and the drawing numbers themselves were part of a centralised system, these were left blank and an apprentice filled them in on *six* copies! Similarly, any modifications to drawings, normally carried out on the original and new prints issued, had to be carried out on the *six* copies, which firstly had to be called in. By this time they had been kicked around the shop and would be filthy and one sacred cow in drawing offices is cleanliness. Then to alter dyelines, the lines to be removed were washed out with potassium permanganate solution which left a brown stain, not only on the drawing but on anything else handy. This stain was then removed with bleach and when the paper was dry, the new version was drawn in on *six* prints.

Standards were very high and only apprentices who had successfully completed drawing office courses at college or university were even considered and sometimes even they were sent packing. Two years' engineering drawing at college amounted to about 80 hours which only represented two weeks' work. Dress standards were rigidly enforced. TIES WERE WORN but in the hottest weather jackets could be removed once the chief gave the O.K. and under a glass roof it could get *very* hot. One African summer student who turned up wearing a patterned orange open-necked shirt outside his trousers, then the avant garde thing, was sacked on the spot.

When my 'time' was out I moved into the top secret project office. All projects were at that time handled by an elderly French designer. His ideas, although well thought out, were rather archaic for the fifties and reflected pre-war European practice. The tendering drawings were somewhat sketchy and showed rigid frame diesel hydraulics with coupling rod drive and 4-8-4. 2-6-2, etc., wheel arrangements. The nearest contemporary thing in appearance was the experimental Fell 4-8-4 diesel-mechanical loco then being tested on British Railways. Just think – all the Western region Warships might have been 4-6-4's had it not been for this man's untimely death. Any bogie locos proposed at this stage had drawgear on the bogies and the bogies coupled together to transmit the tractive effort directly and not through the bogie pivot pins. This line of thought was used on the LNER/BR Manchester-Sheffield electric locos and also our 2-6-6-2 electric locos for South Africa. However it was a false trail causing rough riding and the now conventional position of couplings on the body and pivoted bogies was adopted.

One thing I did learn from this earlier period was a graphical means of evaluating the performance of a proposed locomotive over an undulating road. All that was required was a gradient profile and a horsepower/speed curve and the time taken over the route could be calculated. It even showed up points where the transmission might 'hunt' from one torque ratio to another. I spent one happy week in the office romping up

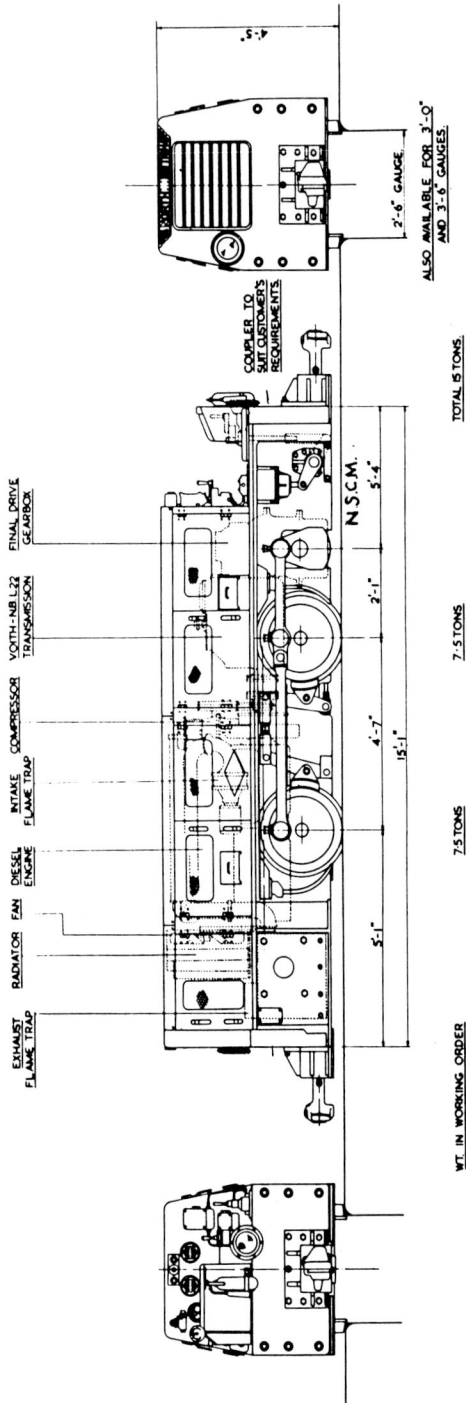
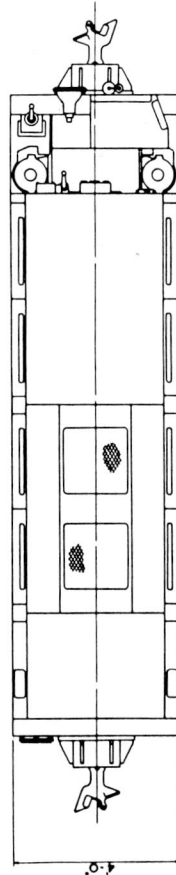

**100BHP Diesel Hydraulic
Mine Locomotive**
North British Locomotive Co. Ltd.
Glasgow

<u>D55-8</u>

68

and down the line from Kuala Lumpur to Singapore with various different powers of engine – a place I've never been to physically, but for a time knew every station.

To build up the diesel section quickly, the NBL had employed several German locomotive engineers on transmission design and development work. I was to work with one of these Germans, one Gunther Ütch who took over the main project design function. Christian names were out with the Germans, so it was Mr Ütch all the time. These men had just come through the complete economic collapse of their country. Germany's miracle of recovery was yet to come, and it showed. Ütch was eating an apple one day, holding it by the stalk and eating it all, core included. When I asked him about this he replied, "Mr Macmillan, if you had lived for a week on a pound of radishes, you would be glad of an apple core."!

Ütch had been involved in railways to help the German war effort. He told of 50 new steam engines which were delivered to the Crimea just before the German retreat began. These were all run over a cliff into the sea but the Russians salvaged every one. A four-metre gauge military railway was proposed from Germany into Russia which could carry tanks transversely on large flat cars. One of his most bizarre accounts was of a plough for tearing up track. This was just that and was pulled behind a locomotive tearing up the sleepers by neatly breaking them in the middle. Unfortunately, with typical German precision, it did the job beautifully and neatly. The Russians used a steam road roller to flatten the sleepers down again and at the same time re-gauging the track to the Russian gauge of 5ft. 0in. A gang of women followed the roller, anchoring the sleepers together with giant staples and the track was replaced almost as quickly as it was torn up. The Russian method of track destruction was much simpler. A piece of rail was bent double round one rail on the track and pulled by a locomotive. This not only tore up the rail but twisted it as well, ensuring the impossibility of relaying it. Another scheme he had worked on was to cope with the slight gauge difference by having wheel sets mounted loose on fixed axles and set at a slight angle. By inverting the axle the gauge could be changed. This fell through as the basic geometry of railway wheels requires the wheels and axles to be rigid with each other.

Utch's main contribution to the British scene was the design of the North British produced bogie locomotives for British Railways. The characteristic curved front windows following the line of roof and bonnet came from his hand. It was not his fault that the diesel electric concept triumphed or that the NB could not produce the M.A.N. diesel engine on steam-age machine tools.

Generally speaking, enquiries for shunters were my province as they were unglamorous and nobody else wished to be bothered with them. It takes as much work to design and tender for one 0-4-0 as for 100 express locos. As it happened we got most of the one- and two-off shunter orders and very few of the big ones. Shunters included everything from 2ft. 0in. gauge underground mine locos to 800 HP 0-8-0's with a design for a sugar plantation 0-6-0 for light narrow gauge work thrown in.

The North British underground mine locos were cuddly little machines only shoulder high and painted yellow but packing a 100 horse power punch. They were 0-4-0's with a jack-shaft drive and coupling rods. Both mechanical or hydraulic transmission could be fitted and they were made to a variety of gauges from 2ft. 0in. to 3ft. 6in. They were fully flameproofed for underground haulage and had washed exhaust gases to prevent pollution. So successful were they that some were supplied with cabs for surface use although that was an expensive solution to a problem. A smaller 75 HP version with cardan shaft drive was planned but only a few were built. One proposal was for two 100 HP miners coupled together and arranged for multiple unit operation. Similar machines are still made by Hunslet who acquired the NBL business through Andrew Barclay's, but the outline has changed and the charisma of the NB Miner has gone.

All enquiries for diesel electrics were tendered as such but a diesel hydraulic alternative offered at slightly more attractive terms. All enquiries for steam shunters were tendered as diesels – "We don't make steam shunting locomotives any more." Barclays usually picked these ones up and did quite nicely out of it. At least they stayed in business.

One project that the entire project office spent weeks working on was a big meaty prestige order for the Coras Impair Eireann, the Irish state railways. O. V. S. Bulleid, late of the Southern Railway in England, was now Chief Mechanical Engineer there and was about to dieselise the lot. The enquiry was for 110 units, 50 at 1600 HP, 50 at 800 HP, and 10 300 HP 0-6-0 shunters. All were to have Bulleid/Firth Brown (BFB) wheels which were his patent and had been used on his S.R. engines and the bogie locos were to be inside framed. Our proposal for the first was a large B--B centre cab diesel having two hydraulic transmissions and full width bodies. The smaller one had one engine and transmission as we had done for India with a metre gauge order but with a cab just offset from centre and both bogies driven by cardan shafts from a centrally mounted gearbox. The shunter would be a standard design adapted for the Irish 5ft. 3in. gauge and using BFB wheels.

I prepared the project drawings and for the big one; six jumper cables were required, four on the centre line and one each side which were handed. These had to be grouped into a cluster on the bonnet and just by chance, their height was such that they formed a configuration

```
     O
   O O O
     O
     O
```

For single unit working a cover was provided so its shape naturally came out thus

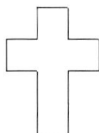

```
    ┌─┐
  ┌─┘ └─┐
  └─┐ ┌─┘
    │ │
    └─┘
```

and the drawings were in a final state before someone spotted the faux pas. We had nearly sent a proposal drawing to Catholic Eire with a crucifx on the front of the engine!

A tremendous effort was put into this tender and when the final documents were made up, special tooled Moroccan leather folders were bought to enclose them with the CIE and NBL's names on them. It was all to no avail. General Motors of America secured the order and their stamp decided the shape of Irish engines to come, not ours. I was sorry because our machines would have had more character, but possibly Ireland was saved the problems BR's Western region were to have.

In the project office one lived on one's nerves. The biggest fright any of us got was when a project became an order. About two weeks were allowed to propose a design. This included determining the configuration of the engine within the customer's specification of weight, power, brakes, etc., estimating the finished weight of the as yet undrawn let alone built engine, estimating the weight of rough material that would be required for the estimating department (non engineers) and getting tenders from all outside suppliers of parts. Only one enquiry in twenty became an order and sub-contractors knew this and also that they would be in competition with others, so they took their time. More often than not, the quoted price for the locomotive had been determined and the tender made before, say, the air brake manufacturers' price had come in. We would 'jump' prices on the basis of previous quotes and inspired guessing. But once we got an order . . . well, you hoped you had got it right. One order for 40 1-Co-Co-1 electric locomotives went sadly wrong. Firstly the finished weights were given to the estimating people instead of the rough weights. Then when the first four locos were delivered, it was found that the frames cracked in service and the power bogie had to be redesigned and 80 new ones built to satisfy the customer. The bogie was a real brute as there was little room to accommodate the large traction motors and keep to South Africa's 3ft. 6in. track gauge. Fortunately for him, the designer responsible for the tender had left the Company's service before all this came out, but we were left to cover the Company's blushes.

One of the NBL's swansongs was to have been a coal burning gas turbine locomotive. Parsons had designed the turbine and it was our job to produce a locomotive for it. The frames were actually built in the Atlas works but it never got any further. The combustion cycle involved burning the coal dust in a heat exchanger and using the heated air to drive the turbine, and the turbine exhaust was combustion air for the coal. The design involved carrying 8 tons of pulverised coal in an open bunker, a rather dangerous practice, and an exhaust temperature that might well have set wooden station awnings and footbridges alight. There is a rather good model of the proposed locomotive in the Glasgow Museum of Transport.

52. NBL 100hp Underground Mine loco adapted for surface use   *(NBL)*

Represented by Units of ...

NORTH BRITISH LOCOMOTIVE Co Ltd

CAPITAL £3,000,000

ISSUED
£750,000 FIVE PER CENT CUMULATIVE PREFERENCE STOCK.
£75,000 FIVE PER CENT 'A' CUMULATIVE PREFERENCE STOCK.
£1,250,000 ORDINARY STOCK.

UNISSUED 175,000 Five per Cent Cumulative Preference Shares of £1 each and 750,000 Ordinary Shares of £1 each.

CERTIFICATE OF ORDINARY STOCK

This is to Certify that   NIGEL S C MACMILLAN

20 Evan Crescent, Giffnock, Glasgow   *is/are the Registered*

Proprietor/s of   ONE   THOUSAND   *Pounds*

of Ordinary Stock as endorsed hereon in the

NORTH BRITISH LOCOMOTIVE Co. Ltd *subject to the Provisions of*

the Memorandum and Articles of Association of the Company

GIVEN *under the* **COMMON SEAL** *of the Company at* **GLASGOW**

this   Eleventh   *day of*   May,   19 61.

71

53. SAR 4-8-2 + 2-8-4 Garratt locomotive outside Hyde Park Works. One of these locomotives has been repatriated and put on show at the Summerlee Museum, Coatbridge. *(Mitchell Library/Springburn Museum)*

Chapter 10

EPILOGUE

Working on tendering, it was fairly obvious that the North British was not landing the kind of orders that spelt prosperity, or even survival. Putting two and two together, I got out from under.

The reasons were complex. The Company was doing its best to serve its customers but the customers were changing. In the classic Indian market, the N B L supplied 100 W G class 2-8-2's in 1950. This was followed by boilers and parts to assemble a further 100 in India and then their own new plant produced ten times that amount without the N B L's help at all.

An impoverished U K Government traded most of the British owned South American railways for meat and the Americans moved in as locomotive suppliers. American aid was widespread elsewhere in the world and the money usually carried the proviso that it had to be spent in the States.

British Railways breathed some life into the order book with their modernisation plan but the N B's contribution was a failure. One contrast between steam and diesel was that whereas with the former the builder did really *build* the entire locomotive, with a diesel he only built the chassis and coachwork. Large electrical manufacturers soon stopped supplying equipment to loco builders and started handling the main contract themselves, it being relatively easy to get chassis built one way or another. The N B L tried manufacturing M.A.N. diesel engines under licence but this was not popular in the UK and combined with the NBL produced hydraulic transmission, the result was a non-standard product that no one wanted.

The axe fell in 1962 and the whole N B L empire collapsed. Ten years later it might have been kept going by massive subsidy, but this philosophy was yet unborn.

The Atlas Works and Hyde Park Works have now been razed to the ground, together with most of the surrounding tenements. The great admin. building is now a technical college, ironically still sporting the words 'Speed' and 'Science' on the original directors' entrance. The stone loco still tops the portal and the N B L's monogram is still on the wrought iron gates. Nearby the new Springburn Museum has been ensconced in the premises of the long established Springburn Library. The museum owns an enormous African Railways 4-8-2+ 2-8-4 Garratt locomotive. Too big by far to be accommodated in Springburn at present (1991) it arrived and was put on display at the Summerlee Heritage Trust in Coatbridge on 29.10.90, being an example of the second last steam order to be completed at the Hyde Park Works in 1956. Poetically, it was hauled into Summerlee behind two steam traction engines.

At Dübs, the boiler shop and machine shop buildings still exist, occupied by general engineering concerns as part of an industrial estate.

For the men who worked there, the training was second to none and most of the professional engineers have prospered in other fields. In the modern world of tape-controlled machine tools, it's quite good to be able to pick up a cold chisel and a 2lb hammer and cut into a piece of steel while the new generation look on – that is, if you can find a chisel in a modern plant.

Beardmore no longer send trainloads of ash to Giffnock quarries. Most of the quarry area has been landscaped as parkland but the course of some of the line can be traced. One enigma still remains: at a far corner of the quarry area and remote from the railway of my childhood, a cast iron chassis with a buffer and safety chain lugs sticks out from the ground under a tree. There is no more reason for this relic to be there than the statues on Easter Island.

The Nitshill line with its primitive wagons has also gone and civilisation in the form of new road layouts and housing is gradually removing all trace.

The East Kilbride branch still flourishes, giving a good service. On 17th April 1965 it was host to a special, hauled by the preserved Highland Railway 4-6-0 and later, an experimental two-car four-wheeled diesel set, prototype for the Pacers, operated for some weeks. Proposals for electrification alternate with closure threats. At present (1991), the all pervading Super Sprinters provide a good service. One interesting idea put forward is to re-open the Clarkston to Muirend loop and run trains to town that way, thus cutting the

54.   The preserved Jones Goods 4-6-0 works a special to East Kilbride. It is passing Overlee playing fields between Clarkston and Busby. The gradient can be appreciated by comparing the level rugby field in the foreground. Easter 1965.

55.   The modern efficient steam locomotives built for South African Railways in the period up to 1958 gave sterling service throughout the 1960's, 70's and 80's, and a few survive in revenue-earning service today. Two 25 class 4-8-4's, the nearest of which is No. 3480 SUSIE (NBL 27340/53), prepare for duty at De Aar on 19 August 1976.   *(Keith Taylorson)*

electrification costs by half. This embankment is still intact, but the one between Clarkston and Williamwood has been bulldozed away leaving only the red bridge over the A726 still standing, making a matching pair with the one carrying the Neilston Electrics.

The Uplawmoor line was cut back to Neilston when electrified but since then has happily carried a fairly intensive service of Glasgow's distinctive electric trains, originally painted a dark blue before the advent of 'rail blue' and now in the Strathclyde livery of flame and black. Even here, the next generation of EMU's is taking over.

Two LNER B1 4-6-0's Nos. 1264 and 1306 MAYFLOWER, and a K1 2-6-0 No. 2005 survive as representatives of NBL-produced UK steam motive power. Nos. 1306 and 2005 have both been restored in LNER apple green livery. The K1 in particular sees active work on BR and has spent several summers working a regular excursion service on the Fort William/Mallaig line, still carrying the author's file marks.

56. Two that got away. B1 No. 1306 steams by at the Shildon cavalcade in 1975.

57. The preserved K1 2-6-0 No. 2009 on the North Yorkshire Moors Railway. Another of the author's efforts.

(W. A. C. Smith)

On this and the following page appear facsimiles of pages from a 24-page booklet "Locomotive Engineering as a Career" published by NBL as a guide to aspiring Apprentices in 1946. NBL were ahead of their time in abolishing Premiums, though the wage rates – from as low as 89p a week for trainees, up to a maximum of only £3.19 – seem absurdly low to present day eyes.

Location of Administrative Offices and Workshops

The Administrative Offices of the Company are situated at 110 Flemington Street, Springburn, Glasgow, N. The Company has three main works: Hyde Park Works and Atlas Works, situated at Springburn, and Queen's Park Works on the southern side of the city.

These extensive works cover a total area of no less than 61 acres, and employ over 5,000 workers.

The Hyde Park and Atlas Works at Springburn, lying on either side of the main London and North Eastern Railway. The Company's Head Offices are shown on the right of this page.

Present-day Outlook for the Locomotive Engineering Industry

This booklet gives an account of the comprehensive training offered to boys who wish to become skilled craftsmen in the Locomotive Industry.

The present-day outlook for the industry is unusually favourable. Not only in Great Britain, but throughout the Empire and the world, the railways are in urgent need of new locomotives.

Thousands of locomotives destroyed during the war have to be replaced, whilst thousands of existing locomotives are due for the scrap-heap and must be renewed.

Moreover, the peacetime programmes of the railways will demand even more " first-line " locomotives than before the war.

There is enough work on hand, on order, and in prospect, to keep the industry in full employment for many years to come.

The North British Locomotive Co. Ltd.

The North British Locomotive Company was formed in 1903, by the amalgamation of some of the oldest firms in the Industry, namely: Sharp, Stewart & Company (previously Sharp, Roberts Co.) established in 1833 ; Neilson Reid & Co. (formerly Neilson & Co.) established in 1837 ; and Dubs & Co., established in 1863.

To-day, outside the United States of America, it is one of the largest locomotive building organizations in the world, with a total output to date of over 26,000 finished locomotives for all parts of the world.

This vast output includes many of the most famous express locomotives of the British Railways, such as " Royal Scot."

During the war, the Company not only maintained a very large production of locomotives, including the new design " Austerity " locomotives, which were shipped to the Continent for the use of the British Liberation Army, but produced, also, tanks, shells, bombs, mines, guns, torpedo tubes and pre-fabricated components for aircraft in very great numbers.

This very fine record of war service is in keeping with the magnificent achievement of the Company in the first great World War.

5

Locomotive Engineering as a Career

Apprenticeship Training Scheme

The Chairman of the Board of Directors of the North British Locomotive Company Limited, Sir Frederick C. Stewart, has stated:

"It is the duty of the Company to see that every apprentice accepted receives the most comprehensive and thorough training in his trade that it is possible to give him. There are two reasons for this: first, for the boy's sake. He deserves the best training we can give him in return for his labour and study. Second, for the Company's sake. We must not forget that the apprentice of to-morrow and, in many cases, the executive foreman of to-day is the craftsman and the Manager."

The scheme outlined on the following pages is designed to put this into effect.

An apprentice forging a locomotive detail with the aid of a steam hammer in one of the Company's forges.

Conditions of Acceptance for Apprenticeship

Physical Fitness

Boys must have a sound physical standard of health to be eligible for training. A medical examination is necessary before a boy can be accepted.

NO PREMIUMS

No premium is payable to any individual or to the Company in consideration or acceptance of an Apprentice.

INDENTURE

As previously stated, all apprentices are indentured for a 5 years' apprenticeship as from the age of 16 or by mutual agreement.

Conditions of Employment

Every Apprentice a production unit.

Every apprentice, both Trade and Engineering, is expected to be a production unit in the Shop.

16

NORTH BRITISH LOCOMOTIVE CO. LTD.

Wages

The rates of pay for apprentices are on a scale in accordance with National Agreements. Wages rise at the end of each year of apprenticeship.

The time spent at classes at the Technical School during working hours is paid for at day rates.

The current rates of pay (June, 1946) for Time-workers are :

1st year	£1	8	1 per week
2nd "	1 13	2	"
3rd "	2 6	0	"
4th "	2 13	7	"
5th "	3 3	10	"

Trainees aged 14 earn 17/10 per week
" " 15 " 23/- "

For those apprentices working on piece work, proportionate increases may be earned.

HOURS OF WORK

The present normal hours of work are :

Monday to Friday   7.45 a.m. to 12.15 p.m.
1.12 p.m. to 5.30 p.m.

Saturdays - - - No normal work.

Apprentices may be required to work overtime.

HOLIDAYS WITH PAY

In common with all other workers employed by the Company, apprentices are entitled to a week's holiday and to 6 other agreed days on full pay every year.

TOOLS, BOOKS, ETC.

Certain small tools of their trade must be provided by the apprentices themselves, but the cost is very low.

Books and Drawing Instruments too, are purchased by apprentices on the advice of the Technical School Authorities, but the total outlay is very reasonable.

CANTEENS

Canteens are available at both Hyde Park and Queen's Park Works for any apprentice who wishes to use them.

Hot mid-day meals are provided at moderate prices.

This apprentice is working inside a locomotive boiler in the Boiler Shop. Boiler making is one of the most interesting branches of locomotive engineering.

Proud of the giant they have helped to build! Apprentices pose for a photograph on one of the powerful new locomotives built by the Company for the L.N.E.R.

Here you see an apprentice turning a locomotive driving wheel. He is gauging the tyre profile with the special gauge you see at top centre.

A book of related interest...
THE RIO TINTO RAILWAY by Alan Sewell

Uniform with this volume, 'The Rio Tinto Railway' is the history of a remarkable British-owned railway that operated for almost 100 years in southern Spain. The 3ft. 6in. gauge railway connected mines at Rio Tinto with the port of Huelva, 51 miles away. A further 99 miles of 2ft. 0in. and 3ft. 6in. gauge track operated within and around the complex of opencast mines, smelters, workshops and power stations at Rio Tinto. The line was powered almost exclusively by British-built locomotives, including numerous examples by Dübs and North British (several of which are pictured at work). Lavishly illustrated with rare and historic pictures (many from the Rio Tinto Co.'s extensive archives, never before published) and maps drawn by the author, this is a book for everyone with an interest in Britain's locomotive and railway heritage, in which 'NBL' played a notable part.

The Rio Tinto Railway (9½in. × 6½in. approx., laminated card covers, 64 pages, 60 b/w photos, 4 maps), is available from Plateway Press, P.O. Box 973, Brighton, BN2 2TG, at £6.95 plus postage.

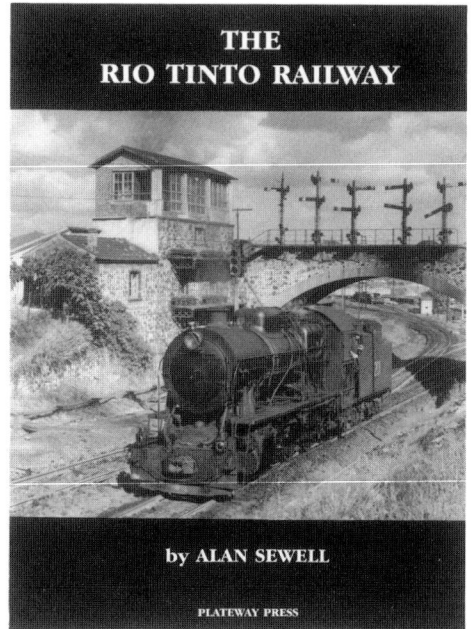

THE RIO TINTO RAILWAY

by ALAN SEWELL

PLATEWAY PRESS

58.   This Dübs 0-6-0T (No. 1894 of 1883) was already nearly 90 years old when photographed at work at Rio Tinto in 1970. Only the Railway's decline and eventual closure prevented it from reaching its centenary – testimony to the longevity of 'NBL' products. (From "The Rio Tinto Railway")
(L. G. Marshall)